STUDY GUIDE
for Stewart, Redlin, and Watson's
COLLEGE ALGEBRA

JOHN A. BANKS
San José State University
San José City College

BROOKS/COLE PUBLISHING COMPANY
Pacific Grove, California

Brooks/Cole Publishing Company
A Division of Wadsworth, Inc.

© 1992 by Wadsworth, Inc., Belmont, California 94002. All rights reserved. This book may not be reproduced, stored in a retrieval system, or transcribed, in any form or by any means—electronic, mechanical, photocopying, recording, or otherwise—without the prior written permission of the publisher, Brooks/Cole Publishing Company, Pacific Grove, California 93950, a division of Wadsworth, Inc.

Printed in the United States of America

10 9 8 7 6 5 4 3 2 1

ISBN 0-534-13004-6

Sponsoring Editor: Faith B. Stoddard
Editorial Assistants: Sarah Wilson and Carol Ann Benedict
Production Coordinator: Dorothy Bell
Cover Design: Katherine Minerva
Printing and Binding: Malloy Lithographing, Inc.

Preface

To master algebra and move on to further studies, a student needs a lot of practice solving problems. The more problems you work on the better your problem-solving skills become. You need to be exposed to many different types of problems and develop strategies to solve each problem.

This study guide provides additional problems for each section of the main text, *College Algebra*, by Stewart, Redlin, and Watson. Each problem includes a detailed solution with commentary on *what* steps are taken and *why* those steps are taken. Some problems are solved using more than one approach, so you can compare the different methods.

It is extremely important that you work out the problems before looking at the solutions. When you immediately look at the solution, before attempting the problem yourself, you short-change yourself. You skip the steps of deciding on the method and planning the strategy you need to solve the problem. These problem-solving skills cannot be developed by looking at the answer first.

Also, students often get accustomed to the order in which the problems appear in a book and learn (memorize) how to solve the problems based on what chapter or section they are in. However, this method will not work on exams where the problems are mixed.

Another method you can use to help you study is using index cards. Put a problem that you find challenging on one side of the card and the solution and location in the text it came from on the other side. Add to your collection whenever you do your homework. Once a week test yourself with these cards. Be sure to shuffle the deck each time.

This study guide was prepared on a Macintosh IIci and printed on an Apple Personal NT. The text was prepared using Word 4.0c. The art was generated in TrueBASIC. For the "viewing rectangles," the figures were imported into SuperPaint, then placed into the Word file. The rest of the figures were prepared using Adobe Illustrator 3.

I hope you find this Study Guide helpful in understanding the concepts in College Algebra.

John A. Banks

Contents

Chapter 1 Basic Algebra

- Section 1.1 What Is Algebra? 2
- Section 1.2 Real Numbers 3
- Section 1.3 Integer Exponents 8
- Section 1.4 Radicals and Rational Exponents 12
- Section 1.5 Algebraic Expressions 15
- Section 1.6 Factoring 18
- Section 1.7 Fractional Expressions 22

Chapter 2 Equations and Inequalities

- Section 2.1 Linear Equations 27
- Section 2.2 Problem Solving With Linear Equations 30
- Section 2.3 Quadratic Equations 35
- Section 2.4 Complex Numbers 41
- Section 2.5 Other Equations 44
- Section 2.6 Linear Inequalities 47
- Section 2.7 Other Inequalities 49
- Section 2.8 Absolute Value 53

Chapter 3 Coordinate Geometry

- Section 3.1 The Coordinate Plane 56
- Section 3.2 Equations and Graphs 60
- Section 3.3 Lines 63
- Section 3.4 Some Special Graphs: Parabolas 67
- Section 3.5 More Special Graphs: Ellipses and Hyperbolas 69

Chapter 4 Functions

- Section 4.1 Functions 75
- Section 4.2 Graphs of Functions 78
- Section 4.3 Graphing Calculators and Computers 82
- Section 4.4 Applied Functions 86
- Section 4.5 Transformation of Functions 88
- Section 4.6 Quadratic Functions and Their Extreme Values 91
- Section 4.7 Using Graphing Devices to Find Extreme Values 95
- Section 4.8 Combining Functions 96
- Section 4.9 Composition of Functions 99
- Section 4.10 One-to-One Functions and Their Inverses 101

Chapter 5 Polynomials and Rational Functions

Section 5.1 Dividing Polynomials	105
Section 5.2 Rational Roots	110
Section 5.3 Using Graphing Devices to Solve Polynomial Equations	117
Section 5.4 Irrational Roots	119
Section 5.5 Complex Roots and the Fundamental Theorem of Algebra	121
Section 5.6 Graphing Polynomials	124
Section 5.7 Using Graphing Devices to Graph Polynomials	127
Section 5.8 Rational Functions	130
Section 5.9 Using Graphing Devices to Graph Rational Functions	136

Chapter 6 Exponential and Logarithmic Functions

Section 6.1 Exponential Functions	140
Section 6.2 Application: Exponential Growth and Decay	142
Section 6.3 Logarithmic Functions	145
Section 6.4 Laws of Logarithms	149
Section 6.5 Applications of Logarithms	154

Chapter 7 Systems of Equations and Inequalities

Section 7.1 Pairs of Lines	159
Section 7.2 Systems of Linear Equations	164
Section 7.3 Inconsistent and Dependent Systems	169
Section 7.4 Algebra of Matrices	174
Section 7.5 Inverses of Matrices and Matrix Equations	179
Section 7.6 Determinants and Cramer's Rule	186
Section 7.7 Nonlinear Systems	189
Section 7.8 Systems of Inequalities	194
Section 7.9 Application: Linear Programming	197
Section 7.10 Applications: Partial Fractions	200

Chapter 8 Counting and Probability

Section 8.1 Counting Principles	204
Section 8.2 Permutations	208
Section 8.3 Combinations	210
Section 8.4 Problem Solving With Permutations and Combinations	212
Section 8.5 Probability	215
Section 8.6 The Union of Events	217
Section 8.7 The Intersection of Events	219
Section 8.8 Expected Value	222

Chapter 9 Sequences and Series

Section 9.1 Sequences 225
Section 9.2 Arithmetic and Geometric Sequences 229
Section 9.3 Series 231
Section 9.4 Arithmetic and Geometric Series 234
Section 9.5 Annuities and Installment Buying 236
Section 9.6 Infinite Geometric Series 238
Section 9.7 Mathematical Induction 241
Section 9.8 The Binomial Theorem 243

Chapter 1

Basic Algebra

Section 1.1 What Is Algebra?
Section 1.2 Real Numbers
Section 1.3 Integer Exponents
Section 1.4 Radicals and Rational Exponents
Section 1.5 Algebraic Expressions
Section 1.6 Factoring
Section 1.7 Fractional Expressions

Section 1.1 What Is Algebra?

Key Ideas
A. Discovering patterns.
B. Using formulas to solve problems.

A. Finding patterns is an important skill that is developed by practice.

1. Write down the general principle exemplified by the following list of facts:
   $2 \cdot 4 + 2 \cdot 5 = 2 \cdot (4 + 5)$
   $2 \cdot 3 + 2 \cdot 11 = 2 \cdot (3 + 11)$
   $2 \cdot 1 + 2 \cdot 8 = 2 \cdot (1 + 8)$

   One way to express this is
   $2 \cdot$ (first number) $+ 2 \cdot$ (second number)
   $= 2 \cdot$ (first number + second number)
   or $2a + 2b = 2(a + b)$
   Be careful that you do not express this pattern as
   $ab + ac = a(b + c)$.
   The list of facts only shows this statement for 2 and not for any number in general.

2. Write down the general principle exemplified by the following list of facts:
   $4 + 7 = 7 + 4$
   $6 + 12 = 12 + 6$
   $5 + 3 = 3 + 5$

   first number + second number =
   second number + first number
   or $a + b = b + a$

B. Finding the pattern leads to writing formulas that can then be used to solve problems.

3. Use the distance formula $D = RT$. A VCR plays 12.5 cm of tape per minute.
   a) How much tape is played in 8 minutes?

   $D = RT$
   $D = \left(\dfrac{12.5 \text{ cm}}{\text{minute}}\right)(8 \text{ minutes}) = 100 \text{ cm}$

   b) How much tape is viewed in T minutes?

   $D = RT$
   $D = \left(\dfrac{12.5 \text{ cm}}{\text{minute}}\right)(T \text{ minutes}) = 12.5T \text{ cm}$

   c) A movie is 2 hours 15 minutes long. How much video tape is needed to put the movie on tape?

   2 hours 15 minutes = 135 minutes
   $D = RT$
   $D = \left(\dfrac{12.5 \text{ cm}}{\text{minute}}\right)(135 \text{ minutes}) = 1687.5 \text{ cm}$

Section 1.2   Real Numbers

Key Ideas
A. What is a real number.
B. Properties of real numbers.
C. Set notations.
D. Interval notation.
E. Absolute value.

A. There are many different types of numbers that make up the **real number** system. Some of these special sets are shown below.

Symbol	Name	Set
N	Natural (counting)	$\{1, 2, 3, 4, ...\}$
Z	Integers	$\{..., -2, -1, 0, 1, 2, ...\}$
Q	Rational	$\{r = \frac{p}{q} \mid p, q \text{ are integers}, q \neq 0\}$
R	Reals	numbers that can be represented by a point on a line.

Every natural number is an integer, and every integer is a rational number. Example: $3 = \frac{3}{1} = \frac{6}{2} = ...$ But not every rational number is an integer and not every integer is a natural number. Real numbers that cannot be expressed as a ratio of integers are called **irrational**. Examples: π and $\frac{1}{\pi}$. Every real number has a decimal representation, and a number is rational when its decimal representation repeats.

1. Classify each real number as a *natural*, *integer*, *rational*, or *irrational* number.

   a) 17.312

   Since a decimal that stops has 0 repeating, that is $17.312 = 17.312\overline{0}$, this number is rational.

   b) −18.101001000100001...

   Since there is not a pattern where a portion is repeated, this number is irrational.
   Here the pattern is 0···01 where the number of zeros grows, so no one sequence is repeated.

   c) $-9.1\overline{27}$

   Since 27 repeats, this number is rational.

   d) 1.5765780...

   Since there is not a pattern where a portion is repeated, this number is irrational.

B. The basic properties used in combining real numbers are:

Commutative Laws	$a + b = b + a$	$ab = ba$
Associative Laws	$(a + b) + c = a + (b + c)$	$(ab)c = a(bc)$
Distributive Law	$a(b + c) = ab + ac$	

Real numbers have another important property called **order.** Order is used to compare two real numbers and determine their relative position.

Order	Symbol	Geometrically
a is *less than* b	$a < b$	a lies to the *left* of b
a is *greater than* b	$a > b$	a lies to the *right* of b
a is *less than or equal* b	$a \leq b$	a lies to the *left* of b or *on* b
a is *greater than or equal* b	$a \geq b$	a lies to the *right* of b or *on* b

2. Use the properties of real numbers to write the given expression without parentheses.

   a) $4(3 + m) - 2(4 + 3m)$

   $4(3 + m) - 2(4 + 3m) = 12 + 4m - 8 - 6m$
   $= 4 - 2m$

   Remember to properly distribute the –2 over $(4 + 3m)$.

   b) $(2a + 3)(5a - 2b + c)$

   $(2a + 3)(5a - 2b + c)$
   $= (2a + 3)5a + (2a + 3)(-2b) + (2a + 3)c$
   $= 10a^2 + 15a - 4ab - 6b + 2ac + 3c$

3. State whether the given inequality is true or false.

   a) $-3.1 > -3$ False.

   b) $2 \leq 2$ True.

   c) $15.3 \geq -16.3$ True.

   d) $2 < -3$ False.

4. Write the statement in terms of an inequality.

   a) w is negative. $w < 0$

   b) m is greater than -3. $-3 < m$

   c) k is at least 6. $6 \leq k$

   d) x is at most 7 and greater -2 $-2 < x \leq 7$

C. A *set* is a collection of objects, called an **element**. A capital letter is usually used to denote sets and lower case letters to represent the elements of the set. There are two main ways to write a set. We can **list** all the elements of the set enclosed in { }, *brackets*, or list a few elements and then use ... to represent that the set continues in the same pattern. Or we can **set builder notation**, "{$x \mid x$ has property P}". This is read as "*the set of x such that x has the property P* ". The two key binary operations for sets are called **union** and **intersection**. The *union* of two sets is the set that consists of the elements that are in *either* set. The *intersection* of two sets is the set that consists of the elements in *both* sets. The **empty set**, ∅, is a set that contains no elements.

5. Let A = {2, 4, 6, 8, 10, 12}, B = {3, 6, 9, 12}, and C = {1, 3, 5, 7, 9, 11}.
   a) Find A ∪ B, A ∪ C, and B ∪ C.

   A ∪ B = {2, 3, 4, 6, 8, 9, 10, 12}
   A ∪ C = {1, 2, 3, 4, 5, 6, 7, 8, 9, 10, 11, 12}
   B ∪ C = {1, 3, 5, 6, 7, 9, 11, 12}

   b) Find A ∩ B, A ∩ C, and B ∩ C.

   A ∩ B = {6, 12}
   A ∩ C = ∅
   B ∩ C = {3, 9}

6. Let A = {$x \mid x < 8$}, B = {$x \mid 3 \leq x < 7$}, and C = {$x \mid 5 < x$}.
   a) Find A ∪ B, A ∪ C, and B ∪ C.

   A ∪ B = {$x \mid x < 8$}
   A ∪ C = {$x \mid -\infty < x < \infty$} = **R**
   B ∪ C = {$x \mid 3 \leq x$}

   b) Find A ∩ B, A ∩ C, and B ∩ C.

   A ∩ B = {$x \mid 3 \leq x < 7$}
   A ∩ C = {$x \mid 5 < x < 8$}
   B ∩ C = {$x \mid 5 < x < 7$}

D. Intervals are sets of real numbers that correspond geometrically to line segments. Study the table on page 9 and note the differences between (and [as well as) and].

7. Let T = (3, 6), V = [2, 4), and W = (4, 6].
   a) Find T ∪ V, T ∪ W, and V ∪ W. Graph your results.

   For unions, the solution consists of the numbers that are in either set. So drawing each set on a number line, we have:

   T ∪ V = (3, 6) ∪ [2, 4)

   = [2, 6)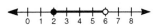

   T ∪ W = (3, 6) ∪ (4, 6]

   = (3, 6]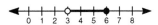

   Note: The only difference between T ∪ W and T is that T ∪ W contains the point 6.

   V ∪ W = [2, 4) ∪ (4, 6]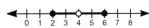

   Note: This is as far as we can go. The sets V and W do not have any elements in common. See V ∩ W below.

   b) Find T ∩ V, T ∩ W, and V ∩ W. Graph your results.

   For intersections, the solution consists of the numbers that are in both intervals.

   T ∩ V = (3, 6) ∩ [2, 4)

   = (3, 4)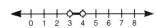

   T ∩ W = (3, 6) ∩ (4, 6]

   = (4, 6)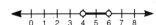

   V ∩ W = ∅

E. The **absolute value** of a number a is the distance from the number a to 0. Remember absolute value is always positive or zero. It is also defined as

$$|a| = \begin{cases} a, & \text{if } a \geq 0 \\ -a, & \text{if } a < 0 \end{cases}$$

8. Find each absolute value.
    a) $|-7|$

    $|-7| = -(-7) = 7$

    b) $|8|$

    $|8| = 8$

    c) $|\pi - 5|$

    $|\pi - 5| = -(\pi - 5) = 5 - \pi$
    since $\pi - 5 < 0$

    d) $|\sqrt{10} - 3|$

    $|\sqrt{10} - 3| = \sqrt{10} - 3$
    Because $\sqrt{10} > \sqrt{9} = 3$,
    $\sqrt{10} - 3 > 0$

9. Find the distance between each pair of numbers.
    a) 4 and 10

    $|4 - 10| = |-6| = 6$
    Remember to do the work on the inside of the absolute value bars *first*!

    b) 5 and –3

    $|5 - (-3)| = |8| = 8$

    c) –2 and –9

    $|-2 - (-9)| = |7| = 7$

Section 1.3 Integer Exponents

Key Ideas
A. Key exponent definitions and rules.
B. Scientific notation.

A. The key exponent definitions are:

$$a^n = a\,a\,a\,\cdots\,a \quad\quad a^0 = 0,\ a \neq 0 \quad\quad a^{-n} = \frac{1}{a^n},\ a \neq 0$$
$$\text{n factors of } a$$

In addition to the exponent definitions, the following key exponent rules should be mastered (memorized).

$$a^m a^n = a^{m+n} \quad\quad \frac{a^m}{a^n} = a^{m-n},\ a \neq 0 \quad\quad (a^m)^n = a^{mn}$$

$$(ab)^n = a^n b^n \quad\quad \left(\frac{a}{b}\right)^n = \frac{a^n}{b^n},\ b \neq 0$$

1. Simplify.

a) $x^3 x^5$ $\quad\quad\quad\quad\quad\quad\quad\quad\quad\quad$ $x^3 x^5 = x^{3+5} = x^8$

b) $w^{12} w^{-8}$ $\quad\quad\quad\quad\quad\quad\quad\quad$ $w^{12} w^{-8} = w^{12-8} = w^4$

c) $\dfrac{y^8}{y^{15}}$ $\quad\quad\quad\quad\quad\quad\quad\quad\quad$ $\dfrac{y^8}{y^{15}} = y^{8-15} = y^{-7} = \dfrac{1}{y^7}$

$\quad\quad\quad\quad\quad\quad\quad\quad\quad\quad\quad\quad$ or $\dfrac{y^8}{y^{15}} = \dfrac{1}{y^{15-8}} = \dfrac{1}{y^7}$

$\quad\quad\quad\quad\quad\quad\quad\quad\quad\quad\quad\quad$ since $\dfrac{1}{a^{-n}} = a^{-(-n)} = a$

d) $(5x)^3$ $\quad\quad\quad\quad\quad\quad\quad\quad\quad$ $(5x)^3 = 5^3 x^3 = 125 x^3$

e) $(m^3)^7$ $\quad\quad\quad\quad\quad\quad\quad\quad$ $(m^3)^7 = m^{3 \cdot 7} = m^{21}$

f) $\dfrac{2^5}{2^7}$

$\dfrac{2^5}{2^7} = 2^{5-7} = 2^{-2} = \dfrac{1}{2^2}$

or $\dfrac{2^5}{2^7} = \dfrac{1}{2^{7-5}} = \dfrac{1}{2^2}$

since $\dfrac{1}{a^{-n}} = a^{-(-n)} = a$

A. Shortcut: In problems like 1c and 1f above, compare the exponents in the numerator and denominator. If the exponent in the numerator is larger, do the work by bringing up the exponent from the denominator. If the exponent in the denominator is larger, do the work by bringing the exponent or the numerator down into the denominator.

2. Simplify $\dfrac{27^3}{9^5}$

Since both 27 and 9 are powers of 3, first express each number as a power of 3.

$\dfrac{27^3}{9^5} = \dfrac{(3^3)^3}{(3^2)^5} = \dfrac{3^{3 \cdot 3}}{3^{2 \cdot 5}} = \dfrac{3^9}{3^{10}} = \dfrac{1}{3^{10-9}} = \dfrac{1}{3}$

3. Simplify $(2x^7y^5)(3x^2y^3)^4$

$(2x^7y^5)(3x^2y^3)^4 = (2x^7y^5)[3^4(x^2)^4(y^3)^4]$
$= (2x^7y^5)(3^4x^8y^{12})$
$= (2)(81)(x^7x^8)(y^5y^{12})$
$= 162x^{15}y^{17}$

4. Simplify $\left(\dfrac{x^4y}{z^6}\right)^3\left(\dfrac{xz^3}{y^4}\right)^5$

$\left(\dfrac{x^4y}{z^6}\right)^3\left(\dfrac{xz^3}{y^4}\right)^5 = \dfrac{(x^4)^3y^3}{(z^6)^3} \cdot \dfrac{x^5(z^3)^5}{(y^4)^5}$

$= \dfrac{x^{12}y^3}{z^{18}} \cdot \dfrac{x^5z^{15}}{y^{20}}$

$= \left(x^4x^5\right)\left(\dfrac{y^3}{y^{20}}\right)\left(\dfrac{z^{15}}{z^{18}}\right)$

$= \dfrac{x^9}{y^{17}z^3}$

5. Eliminate negative exponents and simplify.

a) $\dfrac{3a^3 b^{-4}}{2a^{-2} b^{-1}}$

$\dfrac{3a^3 b^{-4}}{2a^{-2} b^{-1}} = \dfrac{3}{2} a^{3-(-2)} b^{-4-(-1)}$

$= \dfrac{3}{2} a^5 b^{-3}$

$= \dfrac{3a^3}{2b^3}$

b) $\left(\dfrac{4w^3 v^{-4}}{2w^2 v^5}\right)^{-3}$

Method 1: First distribute the –3, then simplify.

$\left(\dfrac{4w^3 v^{-4}}{2w^2 v^5}\right)^{-3} = \dfrac{4^{-3} w^{3(-3)} v^{-4(-3)}}{2^{-3} w^{2(-3)} v^{5(-3)}} = \dfrac{2^{-6} w^{-9} v^{12}}{2^{-3} w^{-6} v^{-15}}$

$= 2^{-6-(-3)} w^{-9-(-6)} v^{12-(-15)}$

$= 2^{-3} w^{-3} v^{27} = \dfrac{v^{27}}{2^3 w^3} = \dfrac{v^{27}}{8w^3}$

Method 2: Simplify inside the parentheses, then distribute the –3.

$\left(\dfrac{4w^3 v^{-4}}{2w^2 v^5}\right)^{-3} = \left(2w^{3-2} v^{-4-5}\right)^{-3} = \left(2wv^{-9}\right)^{-3}$

$= 2^{-3} w^{-3} v^{-9(-3)} = 2^{-3} w^{-3} v^{27}$

$= \dfrac{v^{27}}{2^3 w^3} = \dfrac{v^{27}}{8w^3}$

In addition, other strategies could have been used to solve the problem.

c) $\dfrac{(5^3 m^7 n^2)^3}{(3^4 m^{-2} n^3)^4}$

$\dfrac{(5^3 m^7 n^2)^3}{(3^4 m^{-2} n^3)^4} = \dfrac{5^{3(3)} m^{7(3)} n^{2(3)}}{3^{4(4)} m^{-2(4)} n^{3(4)}}$

$= \dfrac{5^9 m^{21} n^6}{3^{16} m^{-8} n^{12}} = \dfrac{5^9 m^{21-(-8)}}{3^{16} n^{12-6}}$

$= \dfrac{5^9 m^{29}}{3^{16} n^6}$

10

B. Scientific notation is used to express very large or very small numbers in a more convenient way. The goal is to express the number x in the form $a \times 10^n$, where $1 \leq a < 10$ and n is an integer.

6. Write the following numbers in scientific notation.
   a) 23,500,000,000

   2.35×10^{10}

   b) 0.000000000067

   6.7×10^{-11}

7. If $a \approx 4.1 \times 10^{-7}$, $b \approx 1.97 \times 10^9$, and $c \approx 3.24 \times 10^{-3}$ find the quotient $\dfrac{a}{bc}$.

   $\dfrac{a}{bc} \approx \dfrac{4.1 \times 10^{-7}}{(1.97 \times 10^9)(3.24 \times 10^{-3})}$

   $\approx \dfrac{4.1}{(1.97)(3.24)} \times 10^{-7-9-(-3)}$

   $\approx 0.64 \times 10^{-13}$

   $\approx 6.4 \times 10^{-14}$

   Notice that the answer is stated to two significant digits because the least accurate of the numbers has two significant digits.

Section 1.4 Radicals and Rational Exponents

Key Ideas
A. Properties of square roots.
B. Rationalizing the denominator.
C. nth root.
D. Rational Exponents.

A. The **radical** symbol, $\sqrt{}$, means "the positive square root of". So $\sqrt{a} = b$ means $b^2 = a$ where $b \geq 0$. The key rules to master are:
  1. $\sqrt{a^2} = |a|$
  2. If $a \geq 0$ and $b \geq 0$, then $\sqrt{a}\sqrt{b} = \sqrt{ab}$.
  3. If $a \geq 0$ and $b > 0$, then $\sqrt{\dfrac{a}{b}} = \dfrac{\sqrt{a}}{\sqrt{b}}$.

   1. Simplify the following expressions.
      a) $\sqrt{28}$ $\sqrt{28} = \sqrt{4}\sqrt{7} = 2\sqrt{7}$

      b) $\sqrt{50}\sqrt{8}$ $\sqrt{50}\sqrt{8} = \sqrt{50 \cdot 8} = \sqrt{400} = 20$

      c) $\dfrac{\sqrt{54}}{\sqrt{6}}$ $\dfrac{\sqrt{54}}{\sqrt{6}} = \sqrt{\dfrac{54}{6}} = \sqrt{9} = 3$

B. Radicals are eliminated from the denominators by multiplying *both* the numerator and denominator by an appropriate expression. This procedure is called **rationalizing the denominator**. Take advantage of the *difference of squares*, $(a-b)(a+b) = a^2 - b^2$ to clear the denominator when it is a sum or difference of radicals. The factors $(a - b)$ and $(a + b)$ are **conjugates** of each other.

   2. Rationalize the denominator in each of the following expressions.
      a) $\dfrac{1}{\sqrt{2}}$ $\dfrac{1}{\sqrt{2}} = \dfrac{1}{\sqrt{2}} \dfrac{\sqrt{2}}{\sqrt{2}} = \dfrac{\sqrt{2}}{2}$

b) $\sqrt{\dfrac{7}{3}}$

$\sqrt{\dfrac{7}{3}} = \dfrac{\sqrt{7}}{\sqrt{3}} = \dfrac{\sqrt{7}}{\sqrt{3}}\dfrac{\sqrt{3}}{\sqrt{3}} = \dfrac{\sqrt{21}}{3}$

c) $\dfrac{3}{1-\sqrt{2}}$

$\dfrac{3}{1-\sqrt{2}} = \dfrac{3}{(1-\sqrt{2})}\dfrac{(1+\sqrt{2})}{(1+\sqrt{2})}$

$= \dfrac{3+3\sqrt{2}}{1-2}$

$= \dfrac{3+3\sqrt{2}}{-1}$

$= -3-3\sqrt{2}$

d) $\dfrac{\sqrt{3}}{\sqrt{5}+\sqrt{3}}$

$\dfrac{\sqrt{3}}{\sqrt{5}+\sqrt{3}} = \dfrac{\sqrt{3}}{(\sqrt{5}+\sqrt{3})}\dfrac{(\sqrt{5}-\sqrt{3})}{(\sqrt{5}-\sqrt{3})}$

$= \dfrac{\sqrt{15}-3}{5-3}$

$= \dfrac{\sqrt{15}-3}{2}$

C. $\sqrt[n]{a} = b$ means $b^n = a$ and b is called the nth root of a. When n is even, $a \geq 0$ and $b \geq 0$. Key rules to master are:

$\sqrt[n]{ab} = \sqrt[n]{a}\sqrt[n]{b}$ $\sqrt[n]{\dfrac{a}{b}} = \dfrac{\sqrt[n]{a}}{\sqrt[n]{b}}$

$\sqrt[m]{\sqrt[n]{a}} = \sqrt[mn]{a}$ $\sqrt[n]{a^n} = \begin{cases} a & \text{if } n \text{ is odd} \\ |a| & \text{if } n \text{ is even} \end{cases}$

3. Simplify.

a) $\sqrt[4]{81}$

$\sqrt[4]{81} = \sqrt[4]{3^4} = 3$

b) $\sqrt[5]{-32}$

$\sqrt[5]{-32} = \sqrt[5]{(-2)^5} = -2$

Odd roots of negative numbers are negative.

c) $\sqrt[3]{32}$

$\sqrt[3]{32} = \sqrt[3]{2^5} = \sqrt[3]{2^3}\sqrt[3]{2^2} = 2\sqrt[3]{2^2} = 2\sqrt[3]{4}$

D. The key rules of rational exponents are: $a^{1/n} = \sqrt[n]{a}$ and $a^{m/n} = \left(\sqrt[n]{a}\right)^m = \sqrt[n]{(a^m)}$

4. Simplify.

a) $8^{2/3}$

$8^{2/3} = (\sqrt[3]{8})^2 = 2^2 = 4$

or $8^{2/3} = \sqrt[3]{(8)^2} = \sqrt[3]{64} = 4$

Both yield the same solution. However, most students find the first way easier to do.

b) $9^{3/2}$

$9^{3/2} = (\sqrt{9})^3 = 3^3 = 27$

c) $(8x^3y^9)^{5/3}$

$(8x^3y^9)^{5/3} = 8^{5/3}x^{3(5/3)}y^{9(5/3)}$
$= 2^{3(5/3)}x^{3(5/3)}y^{9(5/3)} = 2^5x^5y^{15}$
$= 32x^5y^{15}$

Section 1.5 Algebraic Expressions

Key Ideas
A. Polynomial.
B. Special product formulas.

A. A **variable** is a letter that can represent any number in a given set of numbers, the **domain**. A **constant** represents a *fixed* number. Algebraic expressions using only addition, subtraction, and multiplication are called **polynomials**. The *general form of a polynomial of* degree n in the variable x is
$$a_n x^n + a_{n-1} x^{n-1} + \cdots + a_2 x^2 + a_1 x + a_0$$
where $a_0, a_1, a_2, \ldots, a_{n-1}, a_n$ are constants with $a_n \neq 0$. A polynomial is the sum of **terms** of the form $a_k x^k$ (called **monomials**) where a_k is a constant and k is a nonnegative integer.

1. Determine the degree of each polynomial.

   a) $5x^3 + 3x^2 - 6$ | Degree 3

   b) $9y - 6y^7 + 200y^5$ | Degree 7

   c) -78 | Degree 0

   d) $\sqrt{5} w^4 + 3w^2 - \frac{3}{4} w^2$ | Degree 4

A. Terms with the *same variables* raised to the *same powers* are called **like terms**. Polynomials are added and subtracted using a process called *combining like terms* which utilizes the Distributive Law.

2. Find the sum:
   $(5x^3 + 3x^2 - 2x + 1) + (x^3 - 6x^2 + 5x - 1)$

   First group like terms, then combine like terms.
   $(5x^3 + 3x^2 - 2x + 1) + (x^3 - 6x^2 + 5x - 1)$
   $= (5x^3 + x^3) + (3x^2 - 6x^2) + (-2x + 5x) + (1 - 1)$
   $= 6x^3 - 3x^2 + 3x$

3. Find the difference:
$(7x^3 - x^2 + 5) - (2x^3 - 6x^2 + 9x - 6)$

First, group like terms, then combine like terms.
$(7x^3 - x^2 + 5) - (2x^3 - 6x^2 + 9x - 6)$
$= (7x^3 - 2x^3) + [-x^2 - (-6x^2)] + (-9x) + [5 - (-6)]$
$= 5x^3 + 5x^2 - 9x + 11$

A. The product of two polynomials is found by repeated use of the Distributive Law.

4. Find the product: $(x^2 - 4xy + 2y^2)(x + 4y)$

Start by treating the first expression as a single number and distribute each term over $(x + 4y)$.
$(x^2 - 4xy + 2y^2)(x + 4y)$
$= (x^2 - 4xy + 2y^2)x + (x^2 - 4xy + 2y^2)4y$
Now distribute x over $(x^2 - 4xy + 2y^2)$ and distribute $4y$ over $(x^2 - 4xy + 2y^2)$.
$= x^3 - 4x^2y + 2xy^2 + 4x^2y - 16xy^2 + 2y^3$
$= x^3 - 14xy^2 + 2y^3$

5. Find the product: $\left(\sqrt{x} - \dfrac{4}{\sqrt{x}}\right)(x + \sqrt{x})$

Use the same methods used for multiplying polynomials.
$\left(\sqrt{x} - \dfrac{4}{\sqrt{x}}\right)(x + \sqrt{x})$
$= \left(\sqrt{x} - \dfrac{4}{\sqrt{x}}\right)x + \left(\sqrt{x} - \dfrac{4}{\sqrt{x}}\right)\sqrt{x}$
$= x\sqrt{x} - \dfrac{4x}{\sqrt{x}} + \sqrt{x}\sqrt{x} - \dfrac{4\sqrt{x}}{\sqrt{x}}$
$= x\sqrt{x} - \dfrac{4x}{\sqrt{x}} + x - 4$
$= x\sqrt{x} - 4\sqrt{x} + x - 4$

B. The following products occur so frequently that it is advisable to memorize them. The first one is used to rationalize the denominator in Section 1.4.

$(a - b)(a + b) = a^2 - b^2$ Difference of Squares
$(a + b)^2 = a^2 + 2ab + b^2$ Perfect Square
$(a - b)^2 = a^2 - 2ab + b^2$ Perfect Square
$(a + b)^3 = a^3 + 3a^2b + 3ab^2 + b^3$ Perfect Cube
$(a - b)^3 = a^3 - 3a^2b + 3ab^2 - b^3$ Perfect Cube

The operation of multiplying algebraic expressions is referred to as *expanding*.

6. Use the special product formulas to find the following products:

a) $(3x - 2y^3)^2$

Perfect square:
$$(3x - 2y^3)^2 = (3x)^2 - 2(3x)(2y^3) + (2y^3)^2$$
$$= 9x^2 - 12xy^3 + 4y^6$$

b) $(w^2 + 2w)^3$

Perfect cube:
$(w^3 + 2w)^3$
$$= (w^2)^3 + 3(w^2)^2(2w) + 3(w^2)(2w)^2 + (2w)^3$$
$$= w^6 + 6w^5 + 12w^4 + 8w^3$$

c) $\left(\sqrt{x} + \frac{1}{\sqrt{x}}\right)\left(\sqrt{x} - \frac{1}{\sqrt{x}}\right)$

Difference of squares:
$$\left(\sqrt{x} + \frac{1}{\sqrt{x}}\right)\left(\sqrt{x} - \frac{1}{\sqrt{x}}\right) = (\sqrt{x})^2 - \left(\frac{1}{\sqrt{x}}\right)^2$$
$$= x - \frac{1}{x}$$

Section 1.6 Factoring

Key Ideas
A. Common factors.
B. Factoring quadratics.
C. Special factoring formulas.
D. Grouping.

A. **Factoring** is the process of writing an algebraic expression (a sum) as a product of simpler ones. *Factoring* is one of the most important tools in algebra. *Factoring* is the opposite operation of *expanding*. The simplest of factoring occurs when each term has a *common factor*.

1. Factor $6x^4 - 2x^7 + 8x^8$

   Since the greatest common factor (gcf) of 6, –2, and 8 is 2 and the gcf of $x^4, x^7,$ and x^8 is x^4, so
   $$46x^4 - 2x^7 + 8x^8 = 2x^4(3 - x^3 + 4x^4)$$

2. Factor $4x^2w^3 - 16x^5w^4 + 28x^6w^6$

   Since the gcf of 4, –16, and 28 is 4; the gcf of $x^2, x^5,$ and x^6 is x^2; and the gcf of $w^3, w^4,$ and w^6 is w^3.
   $$4x^2w^3 - 16x^5w^4 + 28x^6w^6$$
   $$= 4x^2w^3(1 - 4x^3w + 7x^4w^3)$$

B. To factor a **quadratic** polynomial, a second-degree polynomial of the form $x^2 + bx + c$, we first observe that $(x + r)(x + s) = x^2 + (r + s)x + rs$, so that $(r + s) = b$ and $rs = c$. If c is negative, then r and s have difference signs and we look for a *difference* of b. If c is positive, then r and s have the same sign as b, either both negative or both positive, and we look for r and s to *sum* to b.

3. Factor $x^2 - 12x + 20$

   Since c is positive and b is negative, r and s are both negative factors of 20 and their sum is 12.
   factors   1·20   2·10   4·5
   corresponding
   sum   21   12   9
   Therefore, taking $r = -2$ and $s = -10$, we get the factorization:
   $$x^2 - 12x + 20 = (x - 2)(x - 10)$$

18

4. Factor $x^2 + 6x - 16$

Since c is negative, r and s have different signs and we look for a difference of 6.

factors	1·16	2·8	4·4
corresponding difference	15	6	0

Since b is positive, the larger factor is positive, so taking $r = 8$ and $s = -2$, we get the factorization:

$$x^2 + 6x - 16 = (x + 8)(x - 2)$$

B. To factor a *general quadratic* of the form $ax^2 + bx + c$, we look for factors of the form $px + r$ and $qx + s$, so $ax^2 + bx + c = (px + r)(qx + s) = pqx^2 + (ps + qr)x + rs$
Therefore, we must find numbers p, q, r, and s such that $pq = a$; $ps + qr = b$; and $rs = c$. If a is positive, then p and q are positive and the signs of r and s are determine as before by the sign of b and c. If a is negative, first factor out a -1.

5. Factor $3x^2 - 7x + 2$

Since c is positive and b is negative, r and s are both negative.

factors of a	1·3	1·3
factors of c	1·2	2·1
corresponding sums	1+6=7	2+3=5

Since $1 \cdot 1 + 3 \cdot 2 = 7$, the 1's must go into different factors and the 3 and 2 must go into different factors. And we get the factorization:

$$3x^2 - 7x + 2 = (x + 2)(3x + 1)$$

C. Formulas 1–3 are used to factor special quadratics, while formulas 4 and 5 are used on the sum or difference of cubics (third degree polynomial).

1. $a^2 - b^2 = (a - b)(a + b)$ Difference of Squares
2. $a^2 + 2ab + b^2 = (a + b)^2$ Prefect Square
3. $a^2 - 2ab + b^2 = (a - b)^2$ Prefect Square
4. $a^3 - b^3 = (a - b)(a^2 + ab + b^2)$ Difference of Cubes
5. $a^3 + b^3 = (a + b)(a^2 - ab + b^2)$ Sum of Cubes

Notice that formulas 1, 4, and 5 have only *two terms* while formulas 2 and 3 have *three terms* each. When factoring a quadratic with three terms and the first and third terms are perfect squares, check the middle term to see if it is $2ab$, if so, then it is a perfect square.

6. Factor.

a) $x^2 - 8x + 16$

Since x^2 and 16 are perfect squares of x and 4, we check the middle term and find that it works, $-8x = -2(x)(4)$. So:

$$x^2 - 8x + 16 = (x - 4)^2$$

b) $25x^2 - 64$

Since this quadratic is the difference of *two terms*, which are both *perfect squares*, we have

$$25x^2 - 64 = (5x - 8)(5x + 8)$$

c) $9x^2 + 12x + 4$

Again, since $9x^2$ and 4 are perfect squares of $3x$ and 2, we check the middle term and find: $12x = 2(3x)(2)$. Factoring, we get:

$$9x^2 + 12x + 4 = (3x + 2)^2$$

d) $8x^3 - 64$

Here we have the difference of two terms, both perfect cubes. Using the difference of cubes, we get:

$$\begin{aligned} 8x^3 - 64 &= (2x)^3 - (4)^2 \\ &= (2x - 4)\left[(2x)^2 + (2x)(4) + (4)^2\right] \\ &= (2x - 4)(4x^2 + 8x + 16) \end{aligned}$$

Note $(4x^2 + 8x + 16)$ *does not factor any further.*

C. In many cases, factoring techniques are combined with the special factoring formulas.

7. Factor.

a) $8x^4 - 24x^3 + 18x^2$

First factor out the common factor of $2x^2$;

$$\begin{aligned} 8x^4 - 24x^3 + 18x^2 &= 2x^2(4x^2 - 12x + 9) \\ &= 2x^2(2x - 3)^2 \end{aligned}$$

b) $3x^2 - 21$

First factor out the common factor 3, then notice that this is the difference of two terms, since any (positive) number can be written as a perfect square, we factor as:
$$3x^2 - 21 = 3(x^2 - 7) = 3(x - \sqrt{7})(x + \sqrt{7})$$

c) $x^6 + 8$

Since x^6 and 8 are each perfect cubes, we use the sum of cubes to factor:
$$\begin{aligned} x^6 + 8 &= (x^2)^3 + (2)^3 \\ &= (x^2 + 2)\left[(x^2)^2 - (x^2)(2) + (2)^2\right] \\ &= (x^2 + 2)(x^4 - 2x^2 + 4) \\ &= (x^2 + 2)(x^2 - \sqrt{2}x + 2)(x^2 + \sqrt{2}x + 2) \end{aligned}$$

The factoring of $(x^4 - 2x^2 + 4)$ is very difficult. For $x^4 - mx^2 + n$ try $\left(x^2 - ax + \sqrt{n}\right)\left(x^2 + ax + \sqrt{n}\right)$ Multiply out these factors and find the value of a.

D. Polynomials with *four terms* can sometimes be factored by grouping the terms into groups of 1 and 3 terms, or 2 and 2 terms, or 3 and 1 terms.

8. Factor $x^6 + 2x^4 + 2x^2 + 4$.

First group into groups of two, then factor out the common factor.
$$\begin{aligned} x^6 + 2x^4 + 2x^2 + 4 &= (x^6 + 2x^4) + (2x^2 + 4) \\ &= x^4(x^2 + 2) + 2(x^2 + 2) \\ &= (x^4 + 2)(x^2 + 2) \end{aligned}$$
Remember, the sum of squares does not factor

9. Factor $x^4 - x^2 + 6x - 9$.

Grouping into groups of two terms each yields no common factor. However, if we group $-x^2 + 6x - 9$ together and factor out a (-1) we get:
$$\begin{aligned} x^4 - x^2 + 6x - 9 &= x^4 + (-x^2 + 6x - 9) \\ &= x^4 - (x^2 - 6x + 9) \\ &= x^4 - (x - 3)^2 \\ &= \left(x^2 - (x - 3)\right)\left(x^2 + (x - 3)\right) \\ &= (x^2 - x + 3)(x^2 + x - 3) \end{aligned}$$

Section 1.7   Fractional Expressions

Key Ideas
A. Rational expressions, simplifying fractions.
B. Multiplying fractions.
C. Dividing fractions.
D. Adding fractions.

A. A quotient of two algebraic expressions is called a **fractional expression**. We deal only with values of the variables that does *not* make the denominator zero. **Rational expressions** are fractional expressions where the numerator and denominator are both polynomials. We simplify fractional expressions in the same way we simplify common fractions: First factor and then cancel common factors.

1. Simplify $\dfrac{x^2 + 2x - 3}{x^2 - 1}$

$$\dfrac{x^2 + 2x - 3}{x^2 - 1} = \dfrac{(x+3)(x-1)}{(x+1)(x-1)} \quad \text{factor}$$

$$= \dfrac{x+3}{x+1} \quad \text{cancel}$$

B. The rule $\dfrac{a}{b} \cdot \dfrac{c}{d} = \dfrac{a \cdot c}{b \cdot d}$ is used to multiply two fractional expressions. However, before multiplying the numerator and denominator, first factor each and simplify by removing the common factor then multiply.

2. Multiply $\dfrac{x^2 - 2x + 1}{x^2 - 4} \cdot \dfrac{x^2 + 4x + 4}{x^2 + x - 2}$

Factor, cancel common factors.
$$\dfrac{x^2 - 2x + 1}{x^2 - 4} \cdot \dfrac{x^2 + 4x + 4}{x^2 + x - 2}$$

$$= \dfrac{(x-1)(x-1)}{(x+2)(x-2)} \cdot \dfrac{(x+2)(x+2)}{(x+2)(x-1)} \quad \text{factor}$$

$$= \dfrac{x-1}{x-2} \quad \text{cancel}$$

C. The rule $\dfrac{a}{b} \div \dfrac{c}{d} = \dfrac{\frac{a}{b}}{\frac{c}{d}} = \dfrac{a}{b} \cdot \dfrac{d}{c} = \dfrac{a \cdot d}{b \cdot c}$ is used to divide two fractional expressions. This is commonly refered to as "*invert and multiply*".

3. Divide $\dfrac{x^2 - x - 6}{x + 4} \div \dfrac{x^2 - 9}{x^2 + 3x - 4}$

$\dfrac{x^2 - x - 6}{x + 4} \div \dfrac{x^2 - 9}{x^2 + 3x - 4}$

$= \dfrac{x^2 - x - 6}{x + 4} \cdot \dfrac{x^2 + 3x - 4}{x^2 - 9}$

$= \dfrac{(x - 3)(x + 2)}{(x + 4)} \cdot \dfrac{(x + 4)(x - 1)}{(x - 3)(x + 3)}$

$= \dfrac{(x + 2)(x - 1)}{x + 3}$

$= \dfrac{x^2 + x - 2}{x + 3}$

4. Divide $\dfrac{x^3 - 7x^2 - 8x}{x^2 - 2x - 15} \div \dfrac{x^2 - 9x + 8}{x^3 - 9x^2 + 20x}$

$\dfrac{x^3 - 7x^2 - 8x}{x^2 - 2x - 15} \div \dfrac{x^2 - 9x + 8}{x^3 - 9x^2 + 20x}$

$= \dfrac{x^3 - 7x^2 - 8x}{x^2 - 2x - 15} \cdot \dfrac{x^3 - 9x^2 + 20x}{x^2 - 9x + 8}$

$= \dfrac{x(x + 1)(x - 8)}{(x - 5)(x + 3)} \cdot \dfrac{x(x - 4)(x - 5)}{(x - 8)(x - 1)}$

$= \dfrac{x^2(x + 1)(x - 4)}{(x + 3)(x - 1)}$

$= \dfrac{x^4 - 3x^3 - 4x^2}{x^2 + 2x - 3}$

D. Just as in the addition of two rational numbers, two fractional expressions must first have a common denominator before they can be added. Then use $\dfrac{a}{b} + \dfrac{c}{b} = \dfrac{a + c}{b}$.

5. Combine and simplify.
   a) $\dfrac{2}{x - 4} + \dfrac{3x}{4 - x}$

Since $4 - x = -(x - 4)$ and $\dfrac{a}{-b} = -\dfrac{a}{b}$

$\dfrac{2}{x - 4} + \dfrac{3x}{4 - x} = \dfrac{2}{x - 4} + \dfrac{3x}{-(x - 4)}$

$= \dfrac{2}{x - 4} - \dfrac{3x}{x - 4}$

$= \dfrac{2 - 3x}{x - 4}$

b) $\dfrac{2}{x+3} + \dfrac{x}{x+5}$

First get the lcd of $x+3$ and $x+5$ which is $(x+3)(x+5)$

$$\dfrac{2}{x+3} + \dfrac{x}{x+5} = \dfrac{2}{(x+3)}\dfrac{(x+5)}{(x+5)} + \dfrac{x}{(x+5)}\dfrac{(x+3)}{(x+3)}$$

$$= \dfrac{2x+10}{(x+3)(x+5)} + \dfrac{x^2+3x}{(x+3)(x+5)}$$

$$= \dfrac{x^2+5x+10}{(x+3)(x+5)}$$

Notice that the numerator is multiplied out while the denominator is not. Usually you will the use this expression where you will need to know the factors of the denominator.

c) $\dfrac{2}{(x-3)(x+2)} + \dfrac{3}{(x+3)(x+2)}$

Since the lcd of $(x-3)(x+2)$ and $(x+3)(x+2)$ is $(x-3)(x+3)(x+2)$. So

$$\dfrac{2}{(x-3)(x+2)} + \dfrac{3}{(x+3)(x+2)}$$

$$= \dfrac{2}{(x-3)(x+2)}\dfrac{(x+3)}{(x+3)} + \dfrac{3}{(x+3)(x+2)}\dfrac{(x-3)}{(x-3)}$$

$$= \dfrac{2x+6}{(x-3)(x+3)(x+2)} + \dfrac{3x-9}{(x-3)(x+3)(x+2)}$$

$$= \dfrac{5x-3}{(x-3)(x+3)(x+2)}$$

d) $\dfrac{-2}{x^2+4x+4} + \dfrac{1}{x^2+5x+6}$

Start by factoring each denominator. Find the lcd..
$x^2 + 4x + 4 = (x+2)(x+2)$
$x^2 + 5x + 6 = (x+2)(x+3)$
So lcd is $(x+2)(x+2)(x+3)$.

$$\dfrac{-2}{x^2+4x+4} + \dfrac{1}{x^2+5x+6}$$

$$= \dfrac{-2}{(x+2)(x+2)} + \dfrac{1}{(x+2)(x+3)}$$

$$= \dfrac{-2}{(x+2)(x+2)}\dfrac{(x+3)}{(x+3)} + \dfrac{1}{(x+2)(x+3)}\dfrac{(x+2)}{(x+2)}$$

$$= \dfrac{-2x-6}{(x+2)(x+2)(x+3)} + \dfrac{x+2}{(x+2)(x+2)(x+3)}$$

$$= \dfrac{-x-4}{(x+2)(x+2)(x+3)}$$

D. First get a single term in the numerator and a single term denominator of a compound fraction. Then divide.

6. Simplify $\dfrac{\dfrac{y}{x}+1}{\dfrac{x}{y}-\dfrac{y}{x}}$

$\dfrac{\dfrac{y}{x}+1}{\dfrac{x}{y}-\dfrac{y}{x}} = \dfrac{\dfrac{y}{x}+1\left(\dfrac{x}{x}\right)}{\left(\dfrac{x}{y}\right)\left(\dfrac{x}{x}\right)-\left(\dfrac{y}{x}\right)\left(\dfrac{y}{y}\right)}$

$= \dfrac{\dfrac{y}{x}+\dfrac{x}{x}}{\dfrac{x^2}{xy}-\dfrac{y^2}{xy}} = \dfrac{\dfrac{y+x}{x}}{\dfrac{x^2-y^2}{xy}}$

$= \dfrac{(x+y)}{x} \cdot \dfrac{xy}{(x^2-y^2)}$

$= \dfrac{(x+y)}{x} \cdot \dfrac{xy}{(x-y)(x+y)}$

$= \dfrac{y}{x-y}$

7. Simplify $\dfrac{\dfrac{y}{x}+1}{\dfrac{x}{y}-\dfrac{y}{x}}$

Chapter 2

Equations and Inequalities

Section 2.1   Linear Equations
Section 2.2   Problem Solving With Linear Equations
Section 2.3   Quadratic Equations
Section 2.4   Complex Numbers
Section 2.5   Other Equations
Section 2.6   Linear Inequalities
Section 2.7   Other Inequalities
Section 2.8   Absolute Value

Section 2.1 Linear Equations

Key Ideas
A. Solving linear equations.
B. Transforming other equations to a linear equation.

A. A linear equation is a first degree equation. To solve an equation *means* to find all solutions (roots) of the equation. Two equations are **equivalent** if they have *exactly* the same solutions. You go between *equivalent* equations by (1) adding or subtracting the *same* quantity from *both sides* of the equation, or (2) multiplying or dividing *both sides* of the equation by the same nonzero quantity. Always check each solution in the original equation.

1. Solve.

   a) $3x - 9 = 0$

   $3x - 9 = 0$
   $3x = 9$
   $x = 3$
   Check: $3(3) - 9 \stackrel{?}{=} 0$
   $9 - 9 = 0$ ✓

   b) $5x + 7 = x - 1$

   $5x + 7 = x - 1$
   $4x = -8$
   $x = -2$
   Check: $5(-2) + 7 \stackrel{?}{=} (-2) - 1$
   $-10 + 7 \stackrel{?}{=} -3$
   $-3 = -3$ ✓

   c) $x - \dfrac{2}{3} = \dfrac{x}{6}$

   Start by multiplying both sides of the equation by the lcd of both sides, in this case 6.
   $\left(x - \dfrac{2}{3}\right) 6 = \left(\dfrac{x}{6}\right) 6$
   $6x - 4 = x$
   $5x = 4$
   $x = \dfrac{4}{5}$
   Check: $\left(\dfrac{4}{5}\right) - \dfrac{2}{3} \stackrel{?}{=} \dfrac{\left(\dfrac{4}{5}\right)}{6}$
   $\dfrac{12}{15} - \dfrac{10}{15} \stackrel{?}{=} \dfrac{4}{30}$
   $\dfrac{2}{15} \stackrel{?}{=} \dfrac{4}{30}$, yes, checks ✓

B. Rational equations can be transformed to an equivalent linear equation by multiplying both sides by the lcd (provided that the lcd is not zero). In these equations, extraneous solutions can be introduced, so it is extremely important to check your solution in the original equation.

2. Solve.

a) $\dfrac{2x+6}{3x} + \dfrac{1}{2} = \dfrac{1}{6}$

$\dfrac{2x+6}{3x} + \dfrac{1}{2} = \dfrac{1}{6}$

$\left(\dfrac{2x+6}{3x} + \dfrac{1}{2}\right)6x = \left(\dfrac{1}{6}\right)6x$

$\left(\dfrac{2x+6}{3x}\right)6x + \left(\dfrac{1}{2}\right)6x = x$

$4x + 12 + 3x = x$

$7x + 12 = x$

$6x = -12$

$x = -2$

Check: $\dfrac{2(-2)+6}{3(-2)} + \dfrac{1}{2} \stackrel{?}{=} \dfrac{1}{6}$

$\dfrac{2}{-6} + \dfrac{1}{2} \stackrel{?}{=} \dfrac{1}{6}$

$-\dfrac{2}{6} + \dfrac{3}{6} \stackrel{?}{=} \dfrac{1}{6}$

$\dfrac{1}{6} = \dfrac{1}{6}$ ✓

b) $\dfrac{x}{x-1} = \dfrac{x+3}{x+1}$

$\left(\dfrac{x}{x-1}\right)(x-1)(x+1) = \left(\dfrac{x+3}{x+1}\right)(x-1)(x+1)$

$x(x+1) = (x+3)(x-1)$

$x^2 + x = x^2 + 2x - 3$

$-x = -3$

$x = 3$

Check: $\dfrac{(3)}{(3)-1} \stackrel{?}{=} \dfrac{(3)+3}{(3)+1}$

$\dfrac{3}{2} = \dfrac{6}{4}$ ✓

c) $\dfrac{x+2}{x-3} + 2 = \dfrac{10}{2x-6}$

$\dfrac{x+2}{x-3} + 2 = \dfrac{10}{2x-6}$

Since $2x - 6 = 2(x - 3)$

$\left(\dfrac{x+2}{x-3} + 2\right)(2x-6) = \left(\dfrac{10}{2x-6}\right)(2x-6)$

$\left(\dfrac{x+2}{x-3}\right)(2x-6) + 2(2x-6) = \left(\dfrac{10}{2x-6}\right)(2x-6)$

$2x + 4 + 4x - 12 = 10$

$6x - 8 = 10$

$6x = 18$

$x = 3$

Check:

$\dfrac{(3)+2}{(3)-3} + 2 \stackrel{?}{=} \dfrac{10}{2(3)-6}$

$\dfrac{6}{0} + 2 \stackrel{?}{=} \dfrac{10}{0}$

Since the fractions in the last equation are undefined, so there is NO solution.

d) $\dfrac{3x+1}{x-3} + \dfrac{2x+1}{-x+3} = \dfrac{2}{3}$

Since $-x + 3 = -1(x - 3)$, the lcd is $3(x - 3)$

$\dfrac{3x+1}{x-3} + \dfrac{2x+1}{-x+3} = \dfrac{2}{3}$ Multiply *both sides* by the lcd.

$3(x-3)\left(\dfrac{3x+1}{x-3} + \dfrac{2x+1}{-x+3}\right) = 3(x-3)\left(\dfrac{2}{3}\right)$

$3(x-3)\dfrac{(3x+1)}{(x-3)} + 3(x-3)\dfrac{(2x+1)}{-(x-3)} = 2(x-3)$

$3(3x+1) - 3(2x+1) = 2(x-3)$

$9x + 3 - 6x - 3 = 2x - 6$

$3x = 2x - 6$

$x = -6$

Check:

$\dfrac{3(-6)+1}{(-6)-3} + \dfrac{2(-6)+1}{-(-6)+3} \stackrel{?}{=} \dfrac{2}{3}$

$\dfrac{-17}{-9} + \dfrac{-11}{9} \stackrel{?}{=} \dfrac{2}{3}$

$\dfrac{17}{9} - \dfrac{11}{9} = \dfrac{6}{9} = \dfrac{2}{3}$ ✓

Section 2.2 Problem Solving With Linear Equations

Key Ideas
- **A.** Guidelines for solving word problems.
- **B.** Mixture problems.
- **C.** Work problems.

A. The steps involved in solving word problems (1) read the problem; (2) identify the *variable*; (3) express all unknown quantities in terms of the *variable*; (4) relate the quantities; (5) set up an equation; (6) solve the equation; (7) check solutions.

1. A train leaves station A at 7 am and travels north to Chicago at 42 miles per hour. Two hours later, an express train leaves station A and travels north to Chicago at 70 miles per hour. When will the express overtake the slower train?

   After reading through the problem, we find that we are looking for the number of *hours the first train travels*.
   Let t = the hours the first train travels;
   Then $t - 2$ = the hours the second train travels.
   The trains will has traveled the same distance when one overtakes the other, so compare the distance traveled by each train using $D = RT$
   1st train distance = $42\,t$
   2nd train distance = $70\,(t - 2)$
   $$42t = 70(t - 2)$$
   $$42t = 70t - 140$$
   $$-28t = -140$$
   $$t = 5 \text{ hours}$$
   So the express train overtakes the slower train at 12 noon.

   Check:
   First train travels $42(5) = 210$ miles and the second train travels $70(5 - 2) = 70(3) = 210$ miles. √

2. Use the distance formula $D = RT$.
   A bicyclist leaves home and pedals to school at 20 miles per hour. Fifteen minutes later, her roommate leaves home and drives her car to school at 35 miles per hour. When will the car overtake the bicyclist?

   Here the distance travel by the bicyclist = distance traveled by the bicyclist.
   Let t = time the car is drive, in hours.
   so $t + \dfrac{1}{4}$ = time bicyclist rides.
   $$20\left(t + \dfrac{1}{4}\right) = 35t$$
   $$20t + 5 = 35t$$
   $$5 = 15t$$
   $$t = \dfrac{5}{15} = \dfrac{1}{3} \text{ hours or 20 minutes}$$
   Check:
   Bicycles: $20\left(\left(\dfrac{1}{3}\right) + \dfrac{1}{4}\right) = 20\left(\dfrac{7}{12}\right) = \dfrac{35}{3}$ miles
   Car: $35\left(\dfrac{1}{3}\right) = \dfrac{35}{3}$ miles √

3. Wanda has 75 coins, some of the coins are quarters and the rest are dimes. If the total amount is $12.90. How many of each type of coins does she have?

Here we are after the number of quarters and the number of dimes.
Let x = number of quarters
$\underline{75 - x \text{ = number of dimes}}$
75 coins total

Now translate the fact that the value of the coins adds up to $12.90.

$\begin{pmatrix} \text{value of} \\ \text{the quarters} \end{pmatrix} + \begin{pmatrix} \text{value of} \\ \text{the dimes} \end{pmatrix} = 12.90$

$.25x + .10(75 - x) = 12.90$

Solve for x:
$.25x + 7.5 - .10x = 12.90$
$.15x + 7.5 = 12.90$
$.15x = 5.40$
$x = \dfrac{5.40}{.15} = 36$

So Wanda has 36 quarters and $75 - 36 = 39$ dimes

Check:
$.25(36) + .10(39) = 9.00 + 3.90 = 12.90$ ✓

B. Mixture problems are solved by relating the value put in to the final value. In these problems, it is helpful to draw pictures and to fill in the table shown below.

	Value in	Value out
types		
%		
amount		
value		

4. A butcher at the Supermarket has 50 pounds of ground meat that contain 36% fat. How many pounds of ground meat containing 15% fat must be added to obtain a mixture that contains 22% fat?

Let x = the number of pounds of 15% fat ground meat added to make the mixture.
In this case, the amount of fat put into the mixture must equal the amount of fat taken out.

	Value IN		Value OUT
Type:	36% fat	15% fat	22% fat
%	36	15	22
pounds	50	x	$50 + x$
Value	.36(50)	.15(x)	.22(50 + x)

$.36(50) + .15(x) = .22(50 + x)$
$36(50) + 15(x) = 22(50 + x)$ *clear decimals by*
$1800 + 15x = 1100 + 22x$ *multiplying by 100*
$700 = 7x$
So $x = 100$ pounds

Check:
The amount of fat put into the mixture is:
$.36(50) + .15(100) = 18 + 15 = 33$ pounds
The amount of fat in the final product is:
$.22(50 + 100) = .22(150) = 33$ pounds ✓

31

5. A winery make a varietal wine that according to the label contains 95% Cabernet grape juice. Due to a valve error too much Zinfandel juice is added to the blend. As a result, the winery ends up with 1500 gallons of wine that now contains 93% Cabernet juice. In order to make the most Cabernet, they plan to remove a portion of the 93% blend and add the remaining 500 gallons of 100% pure Cabernet so as to reach 95% Cabernet juice. How many gallons do they do they need to remove and replace with pure Cabernet juices?

Let x = gallons of 93% Cabernet blend that is removed.
Then $1500 - x$ is the gallons of the 93% blend that remains.
When 500 gallons is added, the resulting blend will have $2000 - x$ gallons.

Type:	Value IN		Value OUT
	Old blend	Pure Cabernet	Resulting blend
%	93%	100%	95%
pounds	$1500 - x$	500	$2000 - x$
Value	$.93(1500 - x)$	$1(500)$	$.95(2000 - x)$

$.93(1500 - x) + 500 = .95(2000 - x)$
$93(1500 - x) + 50000 = 95(2000 - x)$
$139500 - 93x + 50000 = 190000 - 95x$
$189500 - 93x = 190000 - 95x$
$2x = 500$
so $x = 250$ gallons

Check:
If 250 gallons is removed, there will be 1250 gallons of 93% blend and the resulting blend will have 1750 gallons of Cabernet wine.
The gallons of Cabernet juice put into the blend is:
$.93(1250) + 500 = 1162.5 + 500 = 1662.5$ gallons
The gallons of Cabernet juice in the final blend is:
$.95(1750) = 1662.5$ gallons ✓

C. *Work* problems are usually solved by looking at the *amount of a job done per unit of time*, that is, $\frac{job}{time}$.

6. From past experience, Alexandra can wash a car in 25 minutes, while it takes Raymonde 35 minutes to wash a car by hand. How long would it take if they worked together?

Let t = amount of time (in minutes) it takes Alexandra and Raymonde to wash a car when they work together.

Each minute, Alexandra washes $\frac{1}{25}$ of the car and Raymonde washes $\frac{1}{35}$ of the car.

So $\frac{1}{25}t + \frac{1}{35}t = 1$ *Multiply by lcd: 175*

$175\left(\frac{1}{25}t + \frac{1}{35}t\right) = 175$

$7t + 5t = 175$
$12t = 175$ So $t = 14.5833$ minutes

7. When Richard helps Theresa prepare food for a banquet, it takes 4 hours. **When Theresa prepares the same amount of food by herself, it takes 7 hours.** How long would it take Richard to prepare the food by himself?

Let t = the number of hours it takes Richard to prepare the food himself. Then in one hour, Theresa can do $\frac{1}{7}$ of the job and Richard can do $\frac{1}{t}$ of the job. Since it takes them 4 hours if they work together, we have:

$\left(\frac{1}{7} + \frac{1}{t}\right)4 = 1$ Simplify;

$\frac{4}{7} + \frac{4}{t} = 1$ Multiply by lcd;

$4t + 28 = 7t$

$28 = 3t$

So $t = \frac{28}{3}$ hours or 9 hours and 20 minutes.

8. The new copier at the duplicating center can make 100 copies in 4 minutes while the older copier can make 100 copies in 6 minutes. How long will it take for the two copiers to make 2000 copies if they are both working at the same time?

Let t = the number of minutes it takes the copies to make the copies if they are both working. Then *each minute*, the newer machine can make $\frac{100}{4}$ copies, the older machine $\frac{100}{6}$ copies, and together $\frac{2000}{t}$ copies.

So

$\frac{100}{4} + \frac{100}{6} = \frac{2000}{t}$ Multiply both sides by lcd; $12t$

$300t + 200t = 24000$

$500t = 24000$

So $t = \frac{24000}{500} = 48$ minutes.

33

9. Sue Lin can wash a stack of dishes in $\frac{2}{3}$ the time it takes Hui to wash the same stack. When they work together, it takes 12 minutes to was the stack of dishes. How long does it take each person to wash the stack of dishes?

Let m = the minutes it takes Hui to wash the stack; $\frac{2}{3}m$ = the time it takes Sue Lin to wash the stack
Each minute:
Hui washes $\frac{1}{m}$ of the stack and Sue Lin washes $\frac{1}{\frac{2}{3}m} = \frac{3}{2m}$ of the stack; together $\frac{1}{12}$ of the stack is washed. So
$$\frac{1}{m} + \frac{3}{2m} = \frac{1}{12}$$
$$12m\left(\frac{1}{m} + \frac{3}{2m}\right) = 12m\left(\frac{1}{12}\right) \quad \text{Multiply by lcd;}$$
$$12 + 18 = 30 = m \quad \text{Simplify;}$$
30 minutes for Hui and $\frac{2}{3}m = \frac{2}{3}(30) = 20$ minutes for Sue Lin.

Section 2.3 Quadratic Equations

Key Ideas
A. Zero factor property.
B. Completing the square.
C. Quadratic formula.
D. Problem solving with quadratic equations.

A. Quadratic equations are second-degree equations. Some quadratic equations can be solved by first setting one side equal to zero and factoring the quadratic. Then use the **zero factor property**: $ab = 0$ if and if $a = 0$ or $b = 0$.
Refer back to Section 1.6 for a review of factoring.

1. Solve the equation $x^2 + 3x + 2 = 0$

    $x^2 + 3x + 2 = 0$
    $(x + 1)(x + 2) = 0$ factor;
    $x + 1 = 0$ or $x + 2 = 0$ set each factor = 0;
    $x = -1$ $x = -2$ solve

    check:
    $(-1)^2 + 3(-1) + 2 = 1 - 3 + 2 = 0$ ✓
    $(-2)^2 + 3(-2) + 2 = 4 - 6 + 2 = 0$ ✓

2. Solve the equation $x^2 + 12 = 7x$

    $x^2 + 12 = 7x$
    $x^2 - 7x + 12 = 0$ set = 0;
    $(x - 3)(x - 4) = 0$ factor;
    $x - 3 = 0$ or $x - 4 = 0$ set each factor = 0;
    $x = 3$ $x = 4$ solve

    check:
    $(3)^2 - 7(3) + 12 = 9 - 21 + 12 = 0$ ✓
    $(4)^2 - 7(4) + 12 = 16 - 28 + 12 = 0$ ✓

3. Solve the equation
    $x^2 = 2x + 3$

    $x^2 = 2x + 3$
    $x^2 - 2x - 3 = 0$ set = 0;
    $(x - 3)(x + 1) = 0$ factor;
    $x - 3 = 0$ or $x + 1 = 0$ set each factor = 0;
    $x = 3$ $x = -1$ solve

    check:
    $(3)^2 - 2(3) - 3 = 9 - 6 - 3 = 0$ ✓
    $(-1)^2 - 2(-1) - 3 = 1 + 2 - 3 = 0$ ✓

B. Some equations can also be solved by using the form $x^2 = c$, where c is a constant, and taking the square root of both sides. So $x = \pm \sqrt{c}$.

4. Solve the equation $x^2 = 9$.

$x^2 = 9$
$x = \pm 3$, that is $x = 3$ or $x = -3$
Check each answer.

5. Solve the equation $x^2 - 8 = 0$.

$x^2 - 8 = 0$
$x^2 = 8$
$x = \pm\sqrt{8} = \pm 2\sqrt{2}$
Check each answer.

6. Solve the equation $4x^2 = 24$.

$4x^2 = 24$
$x^2 = 6$ ÷ both sides by 4
$x = \pm\sqrt{6}$

or $(2x)^2 = 24$
$2x = \pm\sqrt{24} = \pm 2\sqrt{6}$
So, $x = \pm\sqrt{6}$
Check each answer.

B. The goal in finding the solution to a quadratic equation by **completing the square** is to add the appropriate constant to make a perfect square, $(x + m)^2$ or $(x - m)^2$. Step one is to isolate the terms involving the variable, x^2 and bx, on one side of the equation. Step two is to determine the value needed to add to both sides to complete the square. Since $(x \pm m)^2 = x^2 \pm 2mx + m^2$, the coefficent of the middle term b must equal $2m$, that is, $m = \frac{1}{2}b$; so add $\left(\frac{b}{2}\right)^2$ to both sides.

7. Solve by completing the square.
$x^2 - 6x + 4 = 0$

$x^2 - 6x + 4 = 0$ isolate the terms with x on one
$x^2 - 6x = -4$ side;
$x^2 - 6x + 9 = -4 + 9$ add $\left(\frac{6}{2}\right)^2 = 3^2 = 9$ to *both sides*;
$(x - 3)^2 = 5$ simplify;
$x - 3 = \pm\sqrt{5}$ take the sq. root *both sides*;
$x = 3 \pm \sqrt{5}$

Be sure to check both answers.

8. Solve the equation by completing the square.
$4x^2 = 12x + 1$

$4x^2 = 12x + 1$	
$4x^2 - 12x = 1$	Isolate the terms with x on one side;
$x^2 - 3x = \dfrac{1}{4}$	÷ both sides by 4;
$x^2 - 3x + \dfrac{9}{4} = \dfrac{1}{4} + \dfrac{9}{4}$	Add $\left(\dfrac{3}{2}\right)^2 = \dfrac{9}{4}$; Simplify;
$\left(x - \dfrac{3}{2}\right)^2 = \dfrac{10}{4}$	Take square root of both sides
$x - \dfrac{3}{2} = \dfrac{\pm\sqrt{10}}{2}$	
$x = \dfrac{3}{2} \pm \dfrac{\sqrt{10}}{2} = \dfrac{3 \pm \sqrt{10}}{2}$	
So $x = \dfrac{3 + \sqrt{10}}{2}$ or $x = \dfrac{3 - \sqrt{10}}{2}$	
Be sure to check both answers.	

C. The **quadratic formula**, $x = \dfrac{-b \pm \sqrt{b^2 - 4ac}}{2a}$, gives the roots (solutions) to the quadratic equation $ax^2 + bx + c = 0$. The radical, $\sqrt{b^2 - 4ac}$ is called the **discriminant, D**. When $D > 0$, the quadratic equation has *two real* solutions. When $D = 0$, there is exactly one solution. And when $D < 0$, the quadratic equation has no real solution.

9. Solve each equation by using the quadratic formula.

a) $4x^2 - 4x - 3 = 0$

Setting $a = 4$, $b = -4$, $c = -3$
$$x = \dfrac{-b \pm \sqrt{b^2 - 4ac}}{2a}$$
$$= \dfrac{-(-4) \pm \sqrt{(-4)^2 - 4(4)(-3)}}{2(4)}$$
$$= \dfrac{4 \pm \sqrt{16 + 48}}{8} = \dfrac{4 \pm \sqrt{64}}{8} = \dfrac{4 \pm 8}{8}$$
So $x = \dfrac{4 + 8}{8} = \dfrac{3}{2}$ or $x = \dfrac{4 - 8}{8} = \dfrac{-1}{2}$

Note: This equation factors into $(2x - 3)(2x + 1) = 0$

b) $3x^2 - 7x + 3 = 0$

Setting $a = 3$, $b = -7$, $c = 3$
$$x = \dfrac{-b \pm \sqrt{b^2 - 4ac}}{2a}$$
$$= \dfrac{-(-7) \pm \sqrt{(-7)^2 - 4(3)(3)}}{2(3)}$$
$$= \dfrac{7 \pm \sqrt{49 - 36}}{6} = \dfrac{7 \pm \sqrt{13}}{6}$$
So $x = \dfrac{7 + \sqrt{13}}{6}$ or $x = \dfrac{7 - \sqrt{13}}{6}$

c) $4x^2 + 20x + 25 = 0$

Setting $a = 4, b = 20, c = 25$

$$x = \frac{-b \pm \sqrt{b^2 - 4ac}}{2a}$$

$$x = \frac{-(20) \pm \sqrt{(20)^2 - 4(4)(25)}}{2(4)}$$

$$= \frac{-20 \pm \sqrt{400 - 400}}{8} =$$

$$= \frac{-20 \pm 0}{8} = \frac{-20}{8}$$

So $x = \frac{-5}{2}$.

Since $4x^2 + 20x + 25 = (2x + 5)^2$, this solution is easy to check.

d) $\frac{3}{x-2} - \frac{2}{x+4} = 3$

$\frac{3}{x-2} - \frac{2}{x+4} = 1$ Multiply both sides by lcd;

$(x+4)(x-2)\left(\frac{3}{x-2} - \frac{2}{x+4}\right) = 1(x+4)(x-2)$

$3(x+4) - 2(x-2) = x^2 + 2x - 8$
$3x + 12 - 2x + 4 = x^2 + 2x - 8$
$x + 16 = x^2 + 2x - 8$
$0 = x^2 + x - 24$

Setting $a = 1, b = 1, c = -24$

$$x = \frac{-(1) \pm \sqrt{(1)^2 - 4(1)(-24)}}{2(1)}$$

$$= \frac{-1 \pm \sqrt{1 + 96}}{2} =$$

$$= \frac{-1 \pm \sqrt{97}}{2}$$

Since neither of these solutions make a denominator zero, both solutions are valid. You should check each solution.

D. Applied problems often lead to quadratic equations. Review Section 2.2 for help in setting up equations from word problems.

10. Linda takes 2 hours more than Greg does to service a paper copier. Together they can service the copier in 5 hours. How long does each take to service the copier?

Let t = be the time it takes Greg to service the copier; $t + 2$ = is the time it takes Linda.
Each hour Greg does $\frac{1}{t}$ and Linda does $\frac{1}{t+2}$ of the job, together they can do $\frac{1}{5}$

$\frac{1}{t} + \frac{1}{t+2} = \frac{1}{5}$ Mulitply by lcd $5t(t+2)$;

$5t(t+2)\left(\frac{1}{t} + \frac{1}{t+2}\right) = 5t(t+2)\left(\frac{1}{5}\right)$

$5(t+2) + 5t = t(t+2)$

$5t + 10 + 5t = t^2 + 2t$

$0 = t^2 - 8t - 10$

Setting $a = 1, b = -8, c = -10$

$t = \dfrac{-b \pm \sqrt{b^2 - 4ac}}{2a}$

$= \dfrac{-(-8) \pm \sqrt{(-8)^2 - 4(1)(-10)}}{2(1)} = \dfrac{8 \pm \sqrt{64 + 40}}{2}$

$= \dfrac{8 \pm \sqrt{104}}{2} = \dfrac{8 \pm 2\sqrt{26}}{2} = 4 \pm \sqrt{26}$

Since $4 - \sqrt{26} < 0$, the only solution is $4 + \sqrt{26}$.

11. A tug tows a barge 24 miles up a river at 10 miles per hours and returns down the river at 12 miles per hour. The entire trip took $5\frac{1}{2}$ hours. What is the rate of the river's current?

Let r = the rate of the river's current.
Then $10 - r$ is the *true* rate up river and $12 + r$ is the *true* rate down river. Use the distance formula and $D = RT$ and solve for time, $T = \frac{D}{R}$.

So the time up river is: $\frac{24}{10-r}$ and the time down river is $\frac{24}{12+r}$.

$$\frac{24}{10-r} + \frac{24}{12+r} = \frac{11}{2} \qquad \text{Multiply by lcd;}$$

$$2(10-r)(12+r)\left(\frac{24}{10-r} + \frac{24}{12+r}\right)$$

$$= 2(10-r)(12+r)\left(\frac{11}{2}\right)$$

$48(12+r) + 48(10-r) = 11(10-r)(12+r)$
$576 + 48r + 480 - 48r = 11(120 - 2r - r^2)$
$1056 = 1320 - 22r - 11r^2$
$0 = -11r^2 - 22r + 264 = -11(r^2 + 2r - 24r^2)$
$0 = -11(r-4)(r+6)$

$r - 4 = 0 \qquad \text{or} \qquad r + 6 = 0$
$r = 4 \qquad\qquad\qquad r = -6$

$r = -6$ does not make sence since it requires the river to flow backwards.
Check $r = 4$:
Up river rate is $10 - 4 = 6$, so the trip up river took $\frac{24}{6}$ = 4 hours, while the down river was at $12 + 4 = 16$ and this portion of the trip took $\frac{24}{16}$ hours or $1\frac{1}{2}$ hours.

Section 2.4 Complex Numbers

Key Ideas
A. Arithmetic operations on complex numbers.
B. Dividing complex numbers, the complex conjugate.
C. Square roots of negative numbers.
D. Imaginary roots of quadratic equations.

A. **Complex numbers** are expressions of the form $a + bi$, where a and b are real numbers and i is the **imaginary number** defined by the equation $i^2 = -1$. a is called the **real part** and bi is called the **complex part** of the number. Two complex numbers are equal if their real and complex parts are the same. When $a = 0$, the number bi, $b \neq 0$ is called a **pure imaginary number**.

1. Add or subtract.
   a) $(3 + 4i) - (4 - 5i)$

   $(3 + 4i) - (4 - 5i) = 3 + 4i - 4 + 5i$
   $= (3 - 4) + (4i + 5i)$
   $= -1 + 9i$

   b) $(\sqrt{3} - i) + (1 + \sqrt{3}i)$

   $(\sqrt{3} - i) + (1 + \sqrt{3}i) = \sqrt{3} - i + 1 + \sqrt{3}i$
   $= \sqrt{3} - 1 + (-1 + \sqrt{3})i$

   This is as far as this expression can be simplified.

2. Multiply.
   a) $(2 - 3i)(7 - 5i)$

   Repeated use of distributive property.
   $(2 - 3i)(7 - 5i) = 2(7 - 5i) + -(3i)(7 - 5i)$
   $= 14 - 10i - 21i + 15i^2$
   $= 14 - 31i - 15$
   $= -1 - 31i$

   b) $(3 + 2i)(4 + 9i)$

   Use the FOIL method
   $(3 + 2i)(4 + 9i) = (3)(4) + (3)(9i) + (2i)(4) + (2i)(9i)$
   $= 12 + 18i + 8i + 18i^2$
   $= 12 + 24i + 18(-1)$
   $= -6 + 24i$

3. Show that $1 + 2i$ is a solution to $x^2 - 2x + 5 = 0$.

   Substitute $1 + 2i$ for x and simplify.
   $(1 + 2i)^2 - 2(1 + 2i) + 5 =$
   $\left((1)^2 + 2(2i)(1) + (2i)^2\right) - 2 - 4i + 5 =$
   $1 + 4i + 4i^2 - 4i + 3 = 1 - 4 + 3 = 0$

   So $1 + 2i$ is a solution.

B. The difference of squares product $(a - bi)(a + bi) = a^2 - (bi)^2 = a^2 - b^2i^2 = a^2 + b^2$ is used to simplify fractional expression or to divide complex numbers. The numbers $a - bi$ and $a + bi$ called **complex conjugates**.

4. Simplify $\dfrac{2}{3-i}$.

$$\dfrac{2}{3-i} = \left(\dfrac{2}{3-i}\right)\left(\dfrac{3+i}{3+i}\right)$$

$$= \dfrac{2(3+i)}{9-i^2} = \dfrac{2(3+i)}{9+1}$$

$$= \dfrac{2(3+i)}{10} = \dfrac{3+i}{5}$$

or $= \dfrac{3}{5} + \dfrac{1}{5}i$

5. Divide $4 - 3i$ by $3 + 2i$.

$$(4-3i) \div (3+2i) = \dfrac{4-3i}{3+2i}$$

$$= \left(\dfrac{4-3i}{3+2i}\right)\left(\dfrac{3-2i}{3-2i}\right)$$

$$= \dfrac{12 - 8i - 9i + 6i^2}{9 - 4i^2}$$

$$= \dfrac{12 - 17i - 6}{9 + 4} = \dfrac{6 - 17i}{13}$$

or $= \dfrac{6}{13} - \dfrac{17}{13}i$.

C. If $-r < 0$, then the square roots of $-r$, $\sqrt{-r}$, are $i\sqrt{r}$ and $-i\sqrt{r}$. $i\sqrt{r}$ is called the **principal square root** of $-r$. Remember $\sqrt{a}\sqrt{b} = \sqrt{ab}$ only when a and b are not both negative. So $\sqrt{-3}\sqrt{-6} = i\sqrt{3}\,i\sqrt{6} = 3\sqrt{2}i^2 = -3\sqrt{2}$, while $\sqrt{(-3)(-6)} = \sqrt{18} = 3\sqrt{2}$.

6. Evaluate.
   a) $\sqrt{-1}$

   $\sqrt{-1} = i$

   b) $\sqrt{-9}$

   $\sqrt{-9} = \sqrt{9}\sqrt{-1} = 3i$

   c) $\sqrt{-12}$

   $\sqrt{-12} = \sqrt{12}\sqrt{-1} = 2\sqrt{3}i$

   d) $-\sqrt{-25}$

   $-\sqrt{-25} = -\sqrt{25}\sqrt{-1}$
   $= -5i$

   Note: $\sqrt{}$ is a grouping symbol, that means "do the work on the inside first".

D. In the quadratic formula, if the *discriminant*, $D = b^2 - 4ac$, is less than 0, then there are no real solutions to $ax^2 + bx + c = 0$, but there will be *two* complex solutions which are conjugates.

7. Solve $x^2 + 4x + 6 = 0$

$x^2 + 4x + 6 = 0$
Setting $a = 1, b = 4, c = 6$ in
$$x = \frac{-b \pm \sqrt{b^2 - 4ac}}{2a}$$
$$x = \frac{-(4) \pm \sqrt{(4)^2 - 4(1)(6)}}{2(1)}$$
$$= \frac{-4 \pm \sqrt{16 - 24}}{2}$$
$$= \frac{-4 \pm \sqrt{-8}}{2} = \frac{-4 \pm 2\sqrt{2}i}{2}$$
$$= \frac{2(-2 \pm \sqrt{2}i)}{2} = \frac{-2 \pm \sqrt{2}i}{2}$$

8. Solve $5x^2 - 6x + 5 = 0$

$5x^2 - 6x + 5 = 0$
Setting $a = 5, b = -6, c = 5$ in
$$x = \frac{-b \pm \sqrt{b^2 - 4ac}}{2a}$$
$$x = \frac{-(-6) \pm \sqrt{(-6)^2 - 4(5)(5)}}{2(5)}$$
$$= \frac{6 \pm \sqrt{36 - 100}}{10}$$
$$= \frac{6 \pm \sqrt{-64}}{10} = \frac{6 \pm 8i}{10}$$
$$= \frac{2(3 \pm 4i)}{2(5)} = \frac{3 \pm 4i}{5}$$

Section 2.5 Other Equations

Key Ideas
A. Polynomial equations.
B. Equations with radicals.

A. Higher degree equations are solved by setting everything equal to zero and factoring. When the variables are all raised to even powers, then the substitution $w = x^2$ can be used.

1. Find all roots of the equation $x^3 - 4x^2 = 5x$.

$x^3 - 4x^2 = 5x$	Set = 0;
$x^3 - 4x^2 - 5x = 0$	Factor;
$x(x^2 - 4x - 5) = 0$	Set each factor = 0;
$x(x-5)(x+1) = 0$	Solve;
$x = 0$ or $x - 5 = 0$ or $x + 1 = 0$	
$x = 0$ or $x = 5$ or $x = -1$	

   The three solutions are $x = 0, 5, -1$. Many students forget the lone x factor and lose the solution $x = 0$. Another common mistake is to divide both sides by x as the first step. This also loses the solution $x = 0$.

2. Find all roots of the equation $x^4 - 3x^2 + 2 = 0$.

   Use the substitution $w = x^2$,
   $x^4 - 3x^2 + 2 = w^2 - 3w + 2$
$w^2 - 3w + 2 = 0$	Factor;
$(w-2)(w-1) = 0$	Set each = 0;
$w - 2 = 0$ or $w - 1 = 0$	Substitute back to x;
$x^2 - 2 = 0$ or $x^2 - 1 = 0$	
$x^2 = 2$ or $x^2 = 1$	
$x = \pm\sqrt{2}$ or $x = \pm 1$	

   Four real solutions are $x = 1, -1, \sqrt{2}, -\sqrt{2}$.

3. Find all roots of the equation $x^4 - 16 = 0$.

   We could use the substitution $w = x^2$, or write as $(x^2)^2 - (4)^2 = 0$, the difference of squares.
   Factoring yields,
   $(x^2 - 4)(x^2 + 4) = 0$
   $(x-2)(x+2)(x^2+4) = 0$
$x - 2 = 0$	$x + 2 = 0$	$x^2 + 4 = 0$
$x = 2$	$x = -2$	$x^2 = -4$
		$x = \pm 2i$

   Two real solutions, $\{-2, 2\}$, two complex solutions $\{-2i, 2i\}$.

44

B. Equations involving one radical are solved by isolating the radical on one side and then *squaring both sides.* When there is more than one radical, put a radical on either side and then square *both sides.* It is very important to check your solutions to these problems. *Extraneous solutions,* incorrect solutions can be introduced into a problem by squaring, for example, $3 \neq -3$ but $(3)^2 = (-3)^2$.

4. Find all real solutions to
   $\sqrt{2x-7} - 5 = x + 4$.

   $\sqrt{2x-7} - 5 = x + 4$ Isolate the $\sqrt{}$;
   $\sqrt{2x-7} = x + 9$ Square *both sides*;
   $2x - 7 = x^2 + 18x + 81$ Simplify;
   $0 = x^2 + 16x + 88$ Set $= 0$;
   $x = \dfrac{-(16) \pm \sqrt{(16)^2 - 4(1)(88)}}{2(1)}$
   $x = \dfrac{-16 \pm \sqrt{256 - 352}}{2}$
   $x = \dfrac{-16 \pm \sqrt{-96}}{2}$
   Since $\sqrt{-96}$ is not a real number, there is no solution to this equation.

5. Find all real solutions to
   $\sqrt{2x} + \sqrt{x+1} = 7$

   $\sqrt{2x} + \sqrt{x+1} = 7$ Move $\sqrt{}$ to each side;
   $\sqrt{2x} = 7 - \sqrt{x+1}$ Square both sides;
   $\left(\sqrt{2x}\right)^2 = \left(7 - \sqrt{x+1}\right)^2$
   $(2x) = 49 - 14\sqrt{x+1} + (x+1)$
   $x - 50 = -14\sqrt{x+1}$ Isolate the $\sqrt{}$;
   $(x-50)^2 = \left(-14\sqrt{x+1}\right)^2$ Square *both sides*;
   $x^2 - 100x + 2500 = 196(x+1)$
   $x^2 - 100x + 2500 = 196x + 196$
   $x^2 - 296x + 2304 = 0$
   $(x - 288)(x - 8) = 0$
   $x - 288 = 0$ $x - 8 = 0$
   $x = 288$ $x = 8$
   Check $x = 288$ *in the original equation*:
   $\sqrt{2(288)} + \sqrt{(288) + 1} = \sqrt{576} + \sqrt{289}$
                         $= 26 + 17 = 43 \neq 7$
   $x = 288$ is not a solution.
   Check $x = 8$ *in the original equation*:
   $\sqrt{2(8)} + \sqrt{(8) + 1} = \sqrt{16} + \sqrt{9}$
                     $= 4 + 3 = 7$
   So, only one solution, $x = 8$.

6. Find all real solutions to
$\sqrt{2x+9} = x+4$

$\sqrt{2x+9} = x+4$
$\left(\sqrt{2x+9}\right)^2 = (x+4)^2$
$2x+9 = x^2 + 8x + 16$
$0 = x^2 + 6x + 7$ Using the quadratic
$x = \dfrac{-6 \pm \sqrt{(6)^2 - 4(1)(7)}}{2(1)}$ formula
$= \dfrac{-6 \pm \sqrt{36-28}}{2} = \dfrac{-6 \pm \sqrt{8}}{2}$
$= \dfrac{-6 \pm 2\sqrt{2}}{2} = \dfrac{2(-3 \pm \sqrt{2})}{2} = -3 \pm \sqrt{2}$

Check $x = -3 + \sqrt{2}$ in the original equation:
$\sqrt{2(-3+\sqrt{2})+9} \stackrel{?}{=} (-3+\sqrt{2}) + 4$
$\sqrt{3+2\sqrt{2}} \stackrel{?}{=} 1+\sqrt{2}$ This is hard to check, but using the definition that $b = \sqrt{a}$ if $b \geq 0$ and $b^2 = a$.
$\left(1+\sqrt{2}\right)^2 = 1 + 2\sqrt{2} + 2 = 3 + 2\sqrt{2}$
So this answer checks.

Check $x = -3 - \sqrt{2}$ in the original equation:
$\sqrt{2(-3-\sqrt{2})+9} \stackrel{?}{=} (-3-\sqrt{2}) + 4$
The right side $(-3-\sqrt{2}) + 4 = 1 - \sqrt{2} < 0$.
Since $\sqrt{2(-3-\sqrt{2})+9} \geq 0$,
this cannot be a solution.

So the only solution is $x = -3 + \sqrt{2}$.

Section 2.6   Linear Inequalities

Key Ideas
A. Rules for manipulating inequalities.
B. Compound inequalities.

A. **Inequalities** are statements involving the symbols $<, >, \leq, \geq$. These rules should be mastered to the point that you know *what* rule to use and *when* and *how* to use it.

1. If $a \leq b$, then $a + c \leq b + c$.   Add/subtract same value from both sides.
2. If $a \leq b$ and $c \leq d$, then $a + c \leq b + d$.
3. If $a \leq b$ and $c > 0$, then $ac \leq bc$.   Multiplying by a *positive* maintains the sign.
4. If $a \leq b$ and $c < 0$, then $ac \geq bc$.   Multiplying by a *negative* FLIPS the sign.
5. If $0 < a < b$, then $\dfrac{1}{a} > \dfrac{1}{b}$

1. Solve and graph the solution on the number line below.
   $4x - 5 < 3$

$4x - 5 < 3$	
$4x < 8$	Add 5 to both sides;
$x < 2$	$\div$ both sides by 4;

2. Solve and graph the solution on the number line below.
   $-\dfrac{1}{3}x + 7 \geq 2x$

$-\dfrac{1}{3}x + 7 \geq 2x$	Add $\dfrac{1}{3}x$ to both sides;
$7 \geq 2x + \dfrac{1}{3}x$	Simplify;
$7 \geq \dfrac{7}{3}x$	Multiply both sides
$3 \geq x$	by $\dfrac{3}{7}$;

3. Solve and graph the solution on the number line below.
   $-8x < 2x - 15$

$-8x < 2x - 15$	Subtract $2x$ from
$-10x < -15$	both sides;
$x > \dfrac{15}{10}$	$\div$ both sides by -10,
$x > \dfrac{3}{2}$	FLIP the sign.

47

B. Compound inequalities use the same rules as simple inequalities. Remember that whatever you do to one part of the inequality you *must* do to the other parts. These inequalities are also usually expressed in increasing order.

4. Solve the inequality $3 \leq 2x - 7 < 8$.

$3 \leq 2x - 7 < 8$
$10 \leq 2x < 15$
$5 \leq x < \dfrac{15}{2}$

add 7;
÷ by 2;

5. Solve the inequality $-7 < 2 - 3x < 11$.

$-7 < 2 - 3x < 11$
$-9 < -3x < 9$
$3 > x > -3$
$-3 < x < 3$

subtract 2;
÷ by –3, FLIP the signs;
Write in increasing order.

6. Solve the inequality $19 < 15 - 4x \leq 27$.

$19 < 15 - 4x \leq 27$
$4 < -4x \leq 12$
$-1 > x \geq -3$
$-3 \leq x < -1$

subtract 15;
÷ by –4, FLIP the signs;
Write in increasing order.

Section 2.7   Other Inequalities

Key Ideas
A. Solving non-linear inequalities.

A. These steps are used to solve inequalities that are not linear.
1. Solve the *equality* to get the points where the inequality *can* change.
2. Determine the points where the inequality is not defined.
3. Use the points from step 1 and step 2 to determine the endpoints of the intervals.
4. Take a test point from within each interval and test it in the original inequality. If the test point satisfies the original inequality, then *every* point in the interval will. If the test point does *not* satisfy the inequality, then *no* point in the interval will.

1. Solve the inequality $x^2 - 2 \leq x$.

> The only places where $x^2 - 2 \leq x$ changes to $x^2 - 2 \geq x$ are where $x^2 - 2 = x$. Solve the equality to find these places.
>
> $x^2 - 2 \leq x$ Solve the equality.
> $x^2 - 2 = x$ Set = 0;
> $x^2 - x - 2 = 0$ Factor;
> $(x - 2)(x + 1) = 0$
> $x = 2$ or $x = -1$
>
> | Intervals | Test pt. | Test |
> |---|---|---|
> | $(-\infty, -1]$ | -2 | $(-2)^2 - 2 \stackrel{?}{\leq} (-2)$ |
> | | | $4 - 2 \stackrel{?}{\leq} -2$, yes |
> | $[-1, 2]$ | 0 | $(0)^2 - 2 \stackrel{?}{\leq} (0)$ |
> | | | $0 - 2 \stackrel{?}{\leq} 0$, no |
> | $[2, \infty)$ | 5 | $(5)^2 - 2 \stackrel{?}{\leq} (5)$ |
> | | | $25 - 2 \stackrel{?}{\leq} 5$, yes |
>
> Solution: $(-\infty, -1] \cup [2, \infty)$

2. Solve the inequality $x^2 - 5x + 6 > 0$

$x^2 - 5x + 6 > 0$
$x^2 - 5x + 6 = 0$ Solve the equality.
$(x - 2)(x - 3) = 0$ Factor;
$x - 2 = 0$ or $x - 3 = 0$
$x = 2$ or $x = 3$

Intervals	Test pt.	Test
$(-\infty, 2]$	0	$(0)^2 - 5(0) + 6 \overset{?}{\leq} 0$
		$6 \overset{?}{\leq} 0$, yes
$[2, 3]$	$\frac{5}{2}$	$\left(\frac{5}{2}\right)^2 - 5\left(\frac{5}{2}\right) + 6 \overset{?}{\leq} 0$
		$\frac{25}{4} - \frac{25}{2} + 6 \overset{?}{\leq} 0$
		$-\frac{25}{4} + 6 \overset{?}{\leq} 0$, no
$[3, \infty)$	5	$(5)^2 - 5(5) + 6 \overset{?}{\leq} 0$
		$25 - 25 + 6 \overset{?}{\leq} 0$, yes

Solution: $(-\infty, -1] \cup [2, \infty)$

3. Solve the inequality $\frac{9}{x} \leq x$

$\frac{9}{x} \leq x$ Solve the equality.
$\frac{9}{x} = \frac{x^2}{x}$ Set = 0;
$\frac{x^2 - 9}{x} = 0$ Set numerator = 0;
$x^2 - 9 = 0$ Solve;
$x = \pm 3$

The statement $\frac{9}{x} \leq x$ is not defined when $x = 0$.

Intervals	Test pt.	Test
$(-\infty, -3]$	-5	$\frac{9}{(-5)} \overset{?}{\leq} (-5)$; no
$[-3, 0)$	-1	$\frac{9}{(-1)} \overset{?}{\leq} (-1)$; yes
		$-9 \overset{?}{\leq} -1$, yes
$(0, 3]$	1	$\frac{9}{(1)} \overset{?}{\leq} (1)$; no
$[3, \infty)$	5	$\frac{9}{(5)} \overset{?}{\leq} (5)$; yes

Solution: $[-3, 0) \cup [3, \infty)$.

Note: The point $x = 0$ is not included because the inequality is not defined there.

4. Solve the inequality $\dfrac{4}{x+1} > \dfrac{3}{x+4}$

$\dfrac{4}{x+1} > \dfrac{3}{x+4}$ Solve the equality.

$\dfrac{4}{x+1} = \dfrac{3}{x+4}$ Set = 0;

$4(x+4) = 3(x+1)$

$4x + 16 = 3x + 3$ Solve;

                  $x = -13$

$\dfrac{4}{x+1} > \dfrac{3}{x+4}$ is not defined when $x = -4$ or $x = -1$.

Intervals	Test pt.	Test
$(-\infty, -13)$	-20	$\dfrac{4}{(-20)+1} \overset{?}{>} \dfrac{3}{(-20)+4}$
		$-\dfrac{4}{19} \overset{?}{>} -\dfrac{3}{16}$; no
$(-13, -4)$	-5	$\dfrac{4}{(-5)+1} \overset{?}{>} \dfrac{3}{(-5)+4}$
		$-\dfrac{4}{4} \overset{?}{>} -\dfrac{3}{1}$; yes
$(-4, -1)$	-3	$\dfrac{4}{(-3)+1} \overset{?}{>} \dfrac{3}{(-3)+4}$
		$\dfrac{4}{-2} \overset{?}{>} \dfrac{3}{1}$; no
$(-1, \infty)$	0	$\dfrac{4}{(0)+1} \overset{?}{>} \dfrac{3}{(0)+4}$
		$4 \overset{?}{>} \dfrac{3}{4}$; yes

Solution: $(-13, -4) \cup (-1, \infty)$.

5. Solve the inequality $\dfrac{2x}{x^2+1} > 0$

$\dfrac{2x}{x^2+1} > 0$ Solve the equality.

$\dfrac{2x}{x^2+1} = 0$ Set = 0;

$2x = 0 \quad x = 0$

$\dfrac{2x}{x^2+1} > 0$

Since $x^2 + 1 \neq 0$, the inequality is defined everywhere,

Intervals	Test pt.	Test
$(-\infty, 0)$	-1	$\dfrac{2(-1)}{(-1)^2+1} \overset{?}{>} 0$
		$\dfrac{-2}{2} \overset{?}{>} 0$; no
$(0, \infty)$	1	$\dfrac{2(1)}{(1)^2+1} \overset{?}{>} 0$
		$\dfrac{2}{2} \overset{?}{>} 0$; yes

Solution: $(0, \infty)$.

6. Solve the inequality $\dfrac{x^2 - 11}{x + 1} \leq 1$

$\dfrac{x^2 - 11}{x + 1} \leq 1$ Solve the equality.

$\dfrac{x^2 - 11}{x + 1} = 1$

$x^2 - 11 = x + 1$

$x^2 - x - 12 = 0$

$(x - 4)(x + 3) = 0$

$x - 4 = 0$ or $x + 3 = 0$

$x = 4 \qquad\quad x = -3$

$\dfrac{x^2 - 11}{x + 1} \leq 1$ is not defined when $x = -1$.

Intervals	Test pt.	Test
$(-\infty, -3]$	-10	$\dfrac{(-10)^2 - 11}{(-10) + 1} \stackrel{?}{\leq} 1$
		$-\dfrac{89}{9} \stackrel{?}{\leq} 1$; yes
$[-3, -1)$	-2	$\dfrac{(-2)^2 - 11}{(-2) + 1} \stackrel{?}{\leq} 1$
		$\dfrac{-7}{-1} \stackrel{?}{\leq} 1$; no
$(-1, 4]$	0	$\dfrac{(0)^2 - 11}{(0) + 1} \stackrel{?}{\leq} 1$
		$-11 \stackrel{?}{\leq} 1$; yes
$[4, \infty)$	5	$\dfrac{(5)^2 - 11}{(5) + 1} \stackrel{?}{\leq} 1$
		$\dfrac{14}{6} \stackrel{?}{\leq} 1$; no

Solution: $(-\infty, -3] \cup (-1, 4]$.

Section 2.8   Absolute Value

Key Ideas
A. Properties of absolute value.
B. Triangle inequality.

A. The following properties of absolute value should be mastered.
1. $|ab| = |a| \, |b|$
2. $\left|\dfrac{a}{b}\right| = \dfrac{|a|}{|b|} \quad b \neq 0$
3. $|a^n| = |a|^n$

   If $a > 0$, then
4. $|x| = a$ if and only if $x = \pm a$
5. $|x| < a$ if and only if $-a < x < a$
6. $|x| > a$ if and only if $x > a$ or $x < -a$

1. Solve $|x + 6| = 8$.

   $x + 6 = 8 \quad \text{or} \quad x + 6 = -8$
   $x = 2 \qquad\qquad\quad x = -14$
   Check:
   $|(2) + 6| = |8| = 8$
   $|(-14) + 6| = |-8| = 8$

2. Solve $|3x + 5| < 7$.

   $|3x + 5| < 7$
   $-7 < 3x + 5 < 7$
   $-12 < 3x < 2$
   $-4 < x < \dfrac{2}{3}$

   Check by looking at x values that are inside the interval and outside the interval.

3. Solve $|5 - 2x| \geq 9$.

   $|5 - 2x| \geq 9$
   $(5 - 2x) \geq 9 \quad \text{or} \quad (5 - 2x) \leq -9$
   $\quad -2x \geq 4 \qquad\qquad\quad -2x \leq -14$
   $\quad\; x \leq -2 \qquad\qquad\qquad\; x \geq 7$

4. Solve $\left|\dfrac{3}{2x+1}\right| \geq 2$

$\left|\dfrac{3}{2x+1}\right| \geq 2$

$\dfrac{|3|}{|2x+1|} \geq 2$

$3 \geq 2|2x+1|$

$|2x+1| \leq \dfrac{3}{2}$ Write in $|\ | \leq a$ form.

$-\dfrac{3}{2} \leq 2x+1 \leq \dfrac{3}{2}$

$-\dfrac{5}{2} \leq 2x \leq \dfrac{1}{2}$

$-\dfrac{5}{4} \leq x \leq \dfrac{1}{4}$

B. A frequently used tool throughout mathematics is the **triangle inequality**. It states:
If a and b are real numbers, then $|a+b| \leq |a| + |b|$

5. Suppose that $|x - 3| < 0.01$ and $|y - 1| < 0.02$
   Show that $|x + y - 4| < 0.03$

   $|x + y - 4| = |x - 3 + y - 1|$
   $\leq |x - 3| + |y - 1|$
   $< 0.01 + 0.02$
   < 0.03

6. Suppose that $|x + 4| < 0.1$, estimate $|x + 6|$.

   $|x + 6| = |x + 4 + 2|$
   $\leq |x + 4| + |2|$
   $< 0.1 + 2 = 2.1$
   So, $|x + 6| < 2.1$

Chapter 3

Coordinate Geometry

Section 3.1 The Coordinate Plane
Section 3.2 Equations and Graphs
Section 3.3 Lines
Section 3.4 Some Special Graphs: Parabolas
Section 3.5 More Special Graphs: Ellipses and Hyperbolas

Section 3.1 The Coordinate Plane

Key Ideas
A. Coordinate system.
B. Distance formula.
C. Midpoint of a line segment.

A. The *x-y* coordinate system is called the **coordinate plane** or the **Cartesian plane**. The points are given as (*x* value, *y* value).

1. Describe each set of points.

   a) $\{(x,y) \mid x = 2\}$.

   $\{(x,y) \mid x = 2\}$.
   $x = 2$ is a vertical line where x is always 2.

   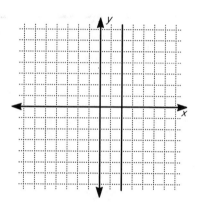

   b) $\{(x,y) \mid \, |y| \geq 3\}$.

   $\{(x,y) \mid \, |y| \geq 3\}$.
   $|y| \geq 3$ is equivalent to
   $y \geq 3$ or $y \leq -3$

   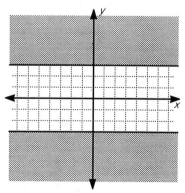

c) $\{(x, y) \mid x \geq 0 \text{ or } y > 0\}$

$\{(x, y) \mid x \geq 0 \text{ or } y > 0\}$

$x \geq 0$ is the y-axis and quadrants I and IV. $y > 0$ is quadrants I and II.

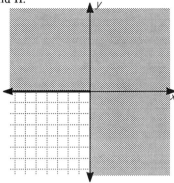

B. The distance between the points $P_1(x_1, y_1)$ and $P_2(x_2, y_2)$ is given by the formula
$|P_1P_2| = \sqrt{(x_2 - x_1)^2 + (y_2 - y_1)^2}$.

2. Find the distance between the following pairs of points.

a) (4, 7) and (5, 5).

$$\begin{aligned} d &= \sqrt{(x_2 - x_1)^2 + (y_2 - y_1)^2} \\ &= \sqrt{(5 - 4)^2 + (5 - 7)^2} \\ &= \sqrt{(1)^2 + (-2)^2} \\ &= \sqrt{1 + 4} \\ &= \sqrt{5} \end{aligned}$$

b) (−2, 3) and (1, −4).

$$\begin{aligned} d &= \sqrt{(x_2 - x_1)^2 + (y_2 - y_1)^2} \\ &= \sqrt{(1 - (-2))^2 + (-4 - 3)^2} \\ &= \sqrt{(3)^2 + (-7)^2} \\ &= \sqrt{9 + 49} \\ &= \sqrt{58} \end{aligned}$$

c) (−5, −1) and (7, −6).

$$\begin{aligned} d &= \sqrt{(x_2 - x_1)^2 + (y_2 - y_1)^2} \\ &= \sqrt{(7 - (-5))^2 + ((-6) - (-1))^2} \\ &= \sqrt{(12)^2 + (-5)^2} \\ &= \sqrt{144 + 25} \\ &= \sqrt{169} = 13 \end{aligned}$$

3. Which point is closer to the origin?
   a) (5, 4) or (–3, –3)?

   Distance from (5, 4) to origin is
   $$= \sqrt{(5-0)^2 + (4-0)^2}$$
   $$= \sqrt{25 + 16}$$
   $$= \sqrt{41}$$
   Distance from (–3, –3) to origin is
   $$= \sqrt{(-3-0)^2 + (-3-0)^2}$$
   $$= \sqrt{9 + 9}$$
   $$= \sqrt{18} = 3\sqrt{2}$$
   Since $\sqrt{18} < \sqrt{41}$, (–3, –3) is closer to the origin.

   b) (7, 2) or (–6, 5)?

   Distance from (7, 2) to origin is
   $$= \sqrt{(7-0)^2 + (2-0)^2}$$
   $$= \sqrt{49 + 4}$$
   $$= \sqrt{53}$$
   Distance from (–6, 5) to origin is
   $$= \sqrt{(-6-0)^2 + (5-0)^2}$$
   $$= \sqrt{36 + 25}$$
   $$= \sqrt{51}$$
   Since $\sqrt{51} < \sqrt{53}$, (–6, 5) is closer to the origin.

C. The **midpoint** of the line segment from $P_1(x_1, y_1)$ to $P_2(x_2, y_2)$ is $\left(\dfrac{x_1 + x_2}{2}, \dfrac{y_1 + y_2}{2}\right)$.

4. Find the midpoint of the line segment from P_1 to P_2.
   a) $P_1 = (1, 5)$ and $P_2 = (-3, -3)$.

   $$\left(\frac{1 + (-3)}{2}, \frac{5 + (-3)}{2}\right) = \left(\frac{-2}{2}, \frac{2}{2}\right)$$
   $$= (-1, 1)$$

   b) $P_1 = (7, -2)$ and $P_2 = (5, 8)$.

   $$\left(\frac{7 + 5}{2}, \frac{(-2) + 8}{2}\right) = \left(\frac{12}{2}, \frac{6}{2}\right)$$
   $$= (6, 3)$$

c) $P_1 = (-5, 4)$ and $P_2 = (-2, 1)$. $\left(\dfrac{(-5) + (-2)}{2}, \dfrac{4+1}{2} \right) = \left(\dfrac{-7}{2}, \dfrac{5}{2} \right)$

Section 3.2   Equations and Graphs

Key Ideas
A. The graph of an equation and Fundamental Principle of Analytic Geometry.
B. Equation of a circle.
C. Symmetry of graphs.

A. The graph of an equation is the set of all points (x, y) that satisfy the equation. The **Fundamental Principle of Analytic Geometry** links Algebra and Geometry. A point (x, y) lies on the graph of an equation if and only if its coordinates satisfy the equation.

1. Graph $x = 2y^2 - 1$.

Since the equation is already solved for x, place in values for y and find the corresponding x value. Make a list.

y	x	(x, y)
-2	7	(7, -2)
-1	1	(1, -1)
0	-1	(-1, 0)
1	1	(1, 1)
2	7	(7, 2)
3	17	(17, 3)

Now plot these points, if you can.

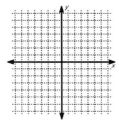

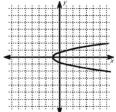

2. Graph $2x + y = 5$.

Start by solving $2x + y = 5$ for y.
$y = -2x + 5$, now put in different values for x and find the corresponding y value.

x	y	(x, y)
-2	9	(-2, 9)
-1	7	(-1, 7)
0	5	(0, 5)
1	3	(1, 3)
2	1	(2, 1)
3	-1	(3, -1)

Now plot these points.

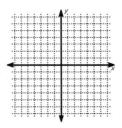

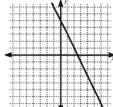

B. The **general equation of a circle** with radius r and centered at (h, k) is:
$(x - h)^2 + (y - k)^2 = r^2$. When the center is the origin, $(0, 0)$, this equation becomes $x^2 + y^2 = r^2$.

3. Determine the graph of $x^2 + y^2 = 49$.

   Writing as $x^2 + y^2 = 7^2$.
   We see that it is a circle of radius 7 centered at the origin.

4. Find an equation of the circle with radius 8 units centered at $(2, -4)$.

   Placing $h = 2$, $k = -4$, and $r = 8$
   $(x - 2)^2 + (y - (-4))^2 = 8^2$
   $(x - 2)^2 + (y + 4)^2 = 64$

5. Determine the center and the radius of the circle given by $x^2 + y^2 + 6x - 4y - 68 = 0$.

   $x^2 + y^2 + 6x - 4y - 68 = 0$ Group, complete the square;
   $x^2 + 6x + \_\_ + y^2 - 4y + \_\_ = 68$
   $x^2 + 6x + 9 + y^2 - 4y + 4 = 68 + 9 + 4$
   $(x + 3)^2 + (y - 2)^2 = 81 = 9^2$

   Center is at $(-3, 2)$, radius is 9 units.

C. Symmetry is an important tool in graphing equations.
   1. A curve is *symmetric with respect to the x-axis* if its equation is unchanged when y is replaced by $-y$.
   2. A curve is *symmetric with respect to the y-axis* if its equation is unchanged when x is replaced by $-x$.
   3. A curve is *symmetric with respect to the origin* if its equation is unchanged when x is replaced by $-x$ and y is replaced by $-y$.

   6. Determine which kind of symmetry the equation $x - y^2 = 6$ has.

      Check each type of symmetry.
      "with respect to the x-axis"
      Yes, since $x - (-y)^2 = x - y^2 = 6$

      "with respect to the y-axis";
      No, $(-x) - y^2 = -x - y^2 \neq x - y^2$

      "with respect to the origin";
      No, $(-x) - (-y)^2 = -x - y^2 \neq x - y^2$

7. Determine which kind of symmetry the equation $x^2 + y^2 - 6y = 2$ has.

Check each type of symmetry.
"with respect to the x-axis"
No, since
$x^2 + (-y)^2 - 6(-y) = x^2 + y^2 + 6y \neq x^2 + y^2 - 6y$
"with respect to the y-axis";
Yes, $(-x)^2 + y^2 - 6y = x^2 + y^2 - 6y = 2$
"with respect to the origin";
No,
$(-x)^2 + (-y)^2 - 6(-y) = x^2 + y^2 + 6y \neq x^2 + y^2 - 6y$

8. Determine which kind of symmetry the equation $x^4 + 3y^2 = 6$ has.

Check each type of symmetry.
"with respect to the x-axis"
Yes, since $x^4 + 3(-y)^2 = x^4 + 3y^2 = 6$
"with respect to the y-axis";
Yes, $(-x)^4 + 3y^2 = x^4 + 3y^2 = 6$
"with respect to the origin";
Yes, $(-x)^4 + 3(-y)^2 = x^4 + 3y^2 = 6$

Section 3.3   Lines

Key Ideas
A. Slope of a line.
B. Point-slope equation.
C. Slope y-intercept equation.
D. Other line equations.
E. Parallel and perpendicular lines.

A. Slope is an extremely important concept in mathematics. Slope is a central focus in the study of calculus. The slope of a nonvertical line that passes through the points $P_1(x_1, y_1)$ and $P_2(x_2, y_2)$ is $m = \dfrac{y_2 - y_1}{x_2 - x_1} = \dfrac{\text{rise}}{\text{run}} = \dfrac{\text{change in } y}{\text{change in } x}$.

1. Find the slope of the line that passes through the points P and Q.

   a) $P(1, 3)$ and $Q(-2, 9)$

   $$m = \frac{9-3}{-2-1} = \frac{6}{-3} = -2$$

   $$m = \frac{3-9}{1-(-2)} = \frac{-6}{3} = -2$$

   The order the points are taken in is not important, but be consistent.

   b) $P(2, -5)$ and $Q(-3, 4)$

   $$m = \frac{4-(-5)}{-3-2} = \frac{9}{-5} = -\frac{9}{5}$$

   c) $P(-7, -4)$ and $Q(-4, -3)$

   $$m = \frac{-3-(-4)}{-4-(-7)} = \frac{1}{3}$$

   d) $P(5, 2)$ and $Q(-3, 2)$

   $$m = \frac{2-2}{-3-5} = \frac{0}{-8} = 0$$

   Note: $\dfrac{0}{8} = 0$ but $\dfrac{8}{0}$ is not defined; many students get these two confused, but they are very different.

e) $P(4, 2)$ and $Q(4, -1)$ $\qquad$ $m = \dfrac{-1-2}{4-4} = \dfrac{-3}{0}$ which is undefined. This is a vertical line.

B. The equation of the **point-slope form** of the line that passes through the point (x_1, y_1) and has slope m is $y - y_1 = m(x - x_1)$. This is a formula you should memorize.

2. Find an equation of the line that passes through the point $(2, -1)$ with slope $\dfrac{1}{3}$. Sketch the line.

    Substituting into $y - y_1 = m(x - x_1)$ with $(x_1, y_1) = (2, -1)$ and $m = \dfrac{1}{3}$.

    $y - (-1) = \dfrac{1}{3}(x - 2)$

    $y + 1 = \dfrac{1}{3}(x - 2)$

    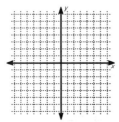
    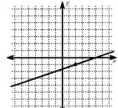

3. Find an equation of the line that passes through the point $(1, 4)$ with slope $\dfrac{-2}{5}$. Sketch the line.

    Substituting into $y - y_1 = m(x - x_1)$ with $(x_1, y_1) = (1, 4)$ and $m = \dfrac{-2}{5}$.

    $y - (4) = \dfrac{-2}{5}(x - 5)$

    $y - 4 = \dfrac{-2}{5}(x - 5)$

    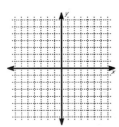
    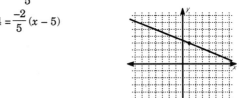

C. The **slope y-intercept equation** of the line, $y = mx + b$, is a special form of the point-slope equation, where the *y-intercept* b is the point $(0, b)$.

4. Find an equation of the line with slope -2 and y-intercept 5.

    Substitute into $y = mx + b$,
    $y = -2x + 5$

5. Find an equation of the line with slope $\frac{4}{5}$ and y-intercept -3.

Substitute into $y = mx + b$,
$y = \frac{4}{5}x - 3$

D. The *general equation* of a line is the equation $ax + by = c$. Every line can be expressed in this form. **Vertical lines** have the form $x = c$ and have *no* slope. **Horizontal lines** have the form $y = d$ and have *zero* slope.

6. Find an equation of the vertical line that passes through the point $(2, -1)$.

$x = 2$

7. Find the slope and y-intercept of the line $3x - 2y = 0$

Solve for y.
$-2y = -3x$
$y = \frac{3}{2}x$
Slope $m = \frac{3}{2}$,
y-intercept $= 0$.

8. Find the slope and y-intercept of the line $2x + 5y = -24$

Solve for y.
$5y = -2x - 24$
$y = \frac{-2x - 24}{5}$
$y = -\frac{2}{5}x - \frac{24}{5}$
Slope $m = -\frac{2}{5}$
y-intercept $= -\frac{24}{5}$.

Remember to divide everything by 5.

$\frac{a+b}{c} = \frac{a}{c} + \frac{b}{c}$

E. Two lines are **parallel** if and only if they have the same slope (or both have *no* slope). Two lines with slope m_1 and m_2 are **perpendicular** if and only if $m_1 m_2 = -1$, that is, their slopes are negative reciprocals. So $m_2 = -\frac{1}{m_1}$. Also, horizontal lines (0 slope) are perpendicular to vertical lines (*no* slope).

9. Determine if each pair of lines are parallel, perpendicular, or neither.

a) $2x + 6y = 7$ and $x + 3y = 9$

Solve for y. Find slope y-intercept form of each line.
$$2x + 6y = 7 \qquad\qquad x + 3y = 9$$
$$6y = -2x + 7 \qquad\qquad 3y = -x + 9$$
$$y = -\frac{1}{3}x + \frac{7}{6} \qquad\qquad y = -\frac{1}{3}x + 3$$
Same slope, $-\frac{1}{3}$, parallel.

b) $5x + 2y = 0$ and $-5x + 2y = 2$

Solve for y. Find slope y-intercept form of each line.
$$5x + 2y = 0 \qquad\qquad -5x + 2y = 2$$
$$2y = -5x \qquad\qquad 2y = 5x + 2$$
$$y = -\frac{5}{2}x \qquad\qquad y = \frac{5}{2}x + 1$$
Neither.

c) $3x - y = 10$ and $x + 3y = 9$

Solve for y. Find slope y-intercept form of each line.
$$3x - y = 10 \qquad\qquad x + 3y = 9$$
$$-y = -3x + 10 \qquad\qquad 3y = -x + 9$$
$$y = 3x - 10 \qquad\qquad y = -\frac{1}{3}x + 3$$
Since $3 = -\dfrac{1}{-\frac{1}{3}}$

10. Are $A(1, 7)$, $B(7, 9)$, $C(9, 3)$, and $D(3, 1)$ the vertices of a rectangle, or parallelogram, or neither.

Sketch the points to find find their relative position. Find the slope of the lines through the pair of points.

slope of $AB = \dfrac{9-7}{7-1} = \dfrac{1}{3}$

slope of $BC = \dfrac{3-9}{9-7} = -3$

slope of $CD = \dfrac{1-3}{3-9} = \dfrac{1}{3}$

slope of $DA = \dfrac{7-1}{1-3} = -3$

Since $AB \parallel CD$ and $BC \parallel DA$, $ABCD$ is at least a parallelogram. Since $AB \perp BC$ this is a rectangle.

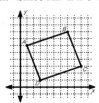

66

Section 3.4   Some Special Graphs: Parabolas

Key Ideas
A. General parabolas.
B. Shifting the vertex.

A. Equation of the form $x = ay^2 + by + c$ or $y = ax^2 + bx + c$ are **parabolas**. The *vertex* is the point where the parabola changes direction. Parabolas of the form $y = ax^2$ opens up when $a > 0$ and opens down when $a < 0$. While parabolas of the form $x = ay^2$ opens to the right when $a > 0$ and opens to the left when $a < 0$. The larger $|a|$ is the wider the parabola is.

1. Graph each pair of parabolas on the same axis. Label each graph.
$y = 2x^2$
$y = \frac{1}{2}x^2$

Generate points on each parabola. Sketch.

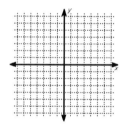
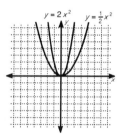

2. Graph each pair of parabolas on the same axis. Label each graph.
$x = y^2$
$x = -2y^2$

Generate points on each parabola. Sketch.

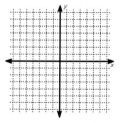
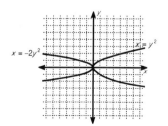

B. The standard equation for a parabola with vertex at (h, k) is either
$x - h = a(y - k)^2$ or $y - k = a(x - h)^2$

3. Find the vertex and sketch the graph of
$x^2 - y = 6$

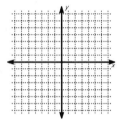

Write in standard form.
$x^2 - y = 6$ Isolate the terms
$x^2 = y + 6$ involving x to one side;

Vertex is at $(0, -6)$, opens up.

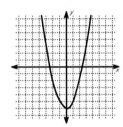

4. Find the vertex and sketch the graph of
$x + 8y = 2y^2 + 6$

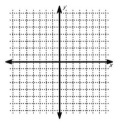

Write in standard form.
$x + 8y = 2y^2 + 6$ Isolate the terms
$x - 6 = 2y^2 - 8y$ involving y to one side;
$x - 6 = 2(y^2 - 4y)$ Factor out the coefficient of y^2;
$x - 6 + 8 = 2(y^2 - 4y + 4)$ Complete the square;
$x + 2 = 2(y - 2)^2$
Vertex is at $(-2, 2)$, opens to the right.

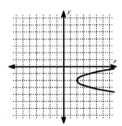

Section 3.5   More Special Graphs: Ellipses and Hyperbolas

Key Ideas
A. Ellipses.
B. Hyperbolas.
C. Standard equation of ellipses and hyperbolas.

A. If $a \neq b$, $a, b > 0$, then the graph of the equation $\dfrac{x^2}{a^2} + \dfrac{y^2}{b^2} = 1$
is an **ellipse** centered at the origin, with x-intercepts at a and $-a$ and y-intercepts at b and $-b$. If $a > b$, then the ellipse is horizontal with *vertices* at $(\pm a, 0)$. The points $(0, \pm b)$ are called *covertices*. If $b > a$, then the ellipse is vertical with vertices at $(0, \pm b)$ and the covvertices are at $(\pm a, 0)$. The line segment joining the vertices is the *major axis* and the line segment joining the covertices is the *minor axis*. The major axis is perpendicular to the minor axis, and they intersect at the *center* of the ellipse.

1. Graph the equation $4x^2 + 25y^2 = 100$.

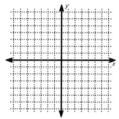

Write in standard form.
$4x^2 + 25y^2 = 100$ ÷ by 100;
$\dfrac{x^2}{25} + \dfrac{y^2}{4} = 1$ or $\dfrac{x^2}{5^2} + \dfrac{y^2}{2^2} = 1$

Ellipse is horizontal with vertices at $(\pm 5, 0)$ and covertices at $(0, \pm 2)$.

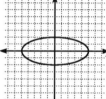

2. Graph the equation $4x^2 + 36y^2 = 9$.

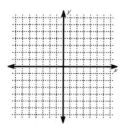

Write in standard form.
$4x^2 + 36y^2 = 9$ ÷ by 9;
$\dfrac{4x^2}{9} + \dfrac{y^2}{4} = 1$
$\dfrac{x^2}{\frac{9}{4}} + \dfrac{y^2}{4} = 1$ or $\dfrac{x^2}{\left(\frac{3}{2}\right)^2} + \dfrac{y^2}{2^2} = 1$

Ellipse is vertical with vertices at $(0, \pm 2)$ and covertices at $\left(\pm \dfrac{3}{2}, 0\right)$.

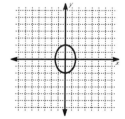

B. If $a \neq b$, $a, b > 0$, then the equations $\dfrac{x^2}{a^2} - \dfrac{y^2}{b^2} = 1$ and $-\dfrac{x^2}{a^2} + \dfrac{y^2}{b^2} = 1$ represent **hyperbolas** centered at the origin, with asymptotes $y = \pm\dfrac{a}{b}x$. The equation $\dfrac{x^2}{a^2} - \dfrac{y^2}{b^2} = 1$ represents a hyperbola with a *horizontal transverse axis* with vertices at $(\pm a, 0)$. This means that one branch of the hyperbola opens to the left while the other branch opens to the right. The equation $-\dfrac{x^2}{a^2} + \dfrac{y^2}{b^2} = 1$ or $\dfrac{y^2}{b^2} - \dfrac{x^2}{a^2} = 1$ represents a hyperbola with a *vertical transverse axis* with vertices at $(0, \pm b)$. This means that one branch of the hyperbola opens up while the other branch opens down.

A shortcut to sketching hyperbolas is to draw a *central box*. The sides of the box are a units to the right and left and b units up and down measured form the center. The diagonals of the central box form the asymptotes of the hyperbola.

3. Graph the equation $8x^2 - 18y^2 = 72$.

Write in standard form.
$8x^2 - 18y^2 = 72$ ÷ by 72;
$\dfrac{x^2}{9} - \dfrac{y^2}{4} = 1$ or $\dfrac{x^2}{3^2} - \dfrac{y^2}{2^2} = 1$

This hyperbola opens horizontally with vertices at $(\pm 3, 0)$.

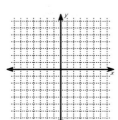

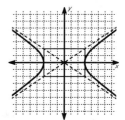

4. Graph the equation $-8x^2 + 18y^2 = 72$.

Write in standard form.
$-8x^2 + 18y^2 = 72$ ÷ by 72;
$-\dfrac{x^2}{9} + \dfrac{y^2}{4} = 1$ or $\dfrac{y^2}{2^2} - \dfrac{x^2}{3^2} = 1$

This hyperbola opens vertically with vertices at $(0, \pm 2)$.

Note: This hyperbola uses the same box as the hyperbola in exercise 3. However, this hyperbola opens up wider.

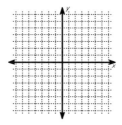

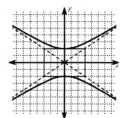

5. Graph the equation $10y^2 - 6x^2 = 4$.

Write in standard form.
$10y^2 - 6x^2 = 4$ ÷ by 4;
$$\frac{5y^2}{2} - \frac{3x^2}{2} = 1$$
$$\frac{y^2}{\left(\frac{2}{5}\right)} - \frac{x^2}{\left(\frac{2}{3}\right)} = 1$$

or $\dfrac{y^2}{\left(\frac{\sqrt{10}}{5}\right)^2} - \dfrac{x^2}{\left(\frac{\sqrt{6}}{3}\right)^2} = 1$

This hyperbola opens vertically with vertices at $\left(0, \pm\dfrac{\sqrt{10}}{5}\right)$

Note: $\sqrt{\dfrac{2}{5}} = \dfrac{\sqrt{2}}{\sqrt{5}} = \dfrac{\sqrt{2}\sqrt{5}}{\sqrt{5}\sqrt{5}} = \dfrac{\sqrt{10}}{5}$

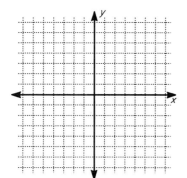
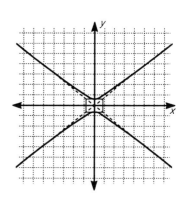

C. Changing x to $x - h$ and y to $y - k$ in an equation shifts the graph of the equation horizontally by h units and vertically by k units. As a result, $\dfrac{(x-h)^2}{a^2} + \dfrac{(y-k)^2}{b^2} = 1$ is the standard equation of an ellipse centered at (h, k). Also, $\dfrac{(x-h)^2}{a^2} - \dfrac{(y-k)^2}{b^2} = 1$ and $\dfrac{(y-k)^2}{b^2} - \dfrac{(x-h)^2}{a^2} = 1$ are the standard equations of the hyperbolas centered at (h, k). For a fixed a, b, h, and k, the three graphs, the ellipse, $\dfrac{(x-h)^2}{a^2} + \dfrac{(y-k)^2}{b^2} = 1$, and the two hyperbolas $\dfrac{(x-h)^2}{a^2} - \dfrac{(y-k)^2}{b^2} = 1$ and $\dfrac{(y-k)^2}{b^2} - \dfrac{(x-h)^2}{a^2} = 1$ are related to the *same* central box. The hyperbolas are graphed on the outside of the box, while the ellipse is graphed on the inside of the box.

6. Determine the type of curve represented by the equation below. Determine the vertices and center.
$4y^2 + 48y + 6x + 131 = x^2$.

Group the variables on one side and the constant on the other.
$4y^2 + 48y - x^2 + 6x = -131$
Group like terms and factor out the coefficient of the squared term.
$(4y^2 + 48y) + (-x^2 + 6x) = -131$
$4(y^2 + 12y + \_\_) - (x^2 - 6x + \_\_) = -131$
Complete the squares.
$4(y^2 + 12y + 36) - (x^2 - 6x + 9) = -131 + 4(36) - 9$
$4(y + 6)^2 - (x - 3)^2 = 4$ ÷ by 4;
$$\frac{(y+6)^2}{1} - \frac{(x-3)^2}{4} = 1$$
$$\frac{(y+6)^2}{1} - \frac{(x-3)^2}{2^2} = 1$$
This is a hyperbola, vertices at $(3, -7)$ and $(3, -5)$ (these are at $(h, k \pm b)$) centered at $(3, -6)$.

7. Determine the type of curve represented by the equation below. Determine the vertices and center.
$4x^2 + 16x + 9y^2 - 90y + 205 = 0$.

Group the variables on one side and the constant on the other.
$4x^2 + 16x + 9y^2 - 90y = -205$
Group like terms and factor out the coefficient of the squared term.
$(4x^2 + 16x) + (9y^2 - 90y) = -205$
$4(x^2 + 4x + \_\_) + 9(y^2 - 10y + \_\_) = -205$
Complete the squares.
$4(x^2 + 4x + 4) + 9(y^2 - 10y + 25) = -205 + 4(4) + 9(25)$
$4(x + 2)^2 + 9(y - 5)^2 = 36$ ÷ by 36;
$$\frac{(x+2)^2}{9} + \frac{(y-5)^2}{4} = 1$$
$$\frac{(x+2)^2}{3^2} + \frac{(y-5)^2}{2^2} = 1$$
This is an ellipse, vertices at $(-5, 5)$ and $(1, 5)$ (these are at $(h \pm a, k)$), centered at $(-2, 5)$.

8. Determine the type of curve represented by the equation below. Determine the vertices and center.
$16y^2 + 32y - 9x^2 + 36x + 124 = 0$.

Group the variables on one side and the constant on the other.
$16y^2 + 32y - 9x^2 + 36x = -124$
Group like terms and factor out the coefficient of the squared term.
$(16y^2 + 32y) + (-9x^2 + 36x) = -124$
$16(y^2 + 2y + \_) - 9(x^2 - 4x + \_) = -124$
Complete the squares.
$16(y^2 + 2y + 1) - 9(x^2 - 4x + 4) = -124 + 16(1) - 9(4)$
$16(y + 1)^2 - 9(x - 2)^2 = -144$
$-\dfrac{(y+1)^2}{9} + \dfrac{(x-2)^2}{16} = 1$
$\dfrac{(x-2)^2}{16} - \dfrac{(y+1)^2}{9} = 1$
$\dfrac{(x-2)^2}{4^2} - \dfrac{(y+1)^2}{3^2} = 1$
This is a hyperbola, vertices at $(0, -1)$ and $(4, -1)$ ((these are at $(h \pm a, k)$), centered at $(2, -1)$.

Chapter 4

Functions

Section 4.1   Functions
Section 4.2   Graphs of Functions
Section 4.3   Graphing Calculators and Computers
Section 4.4   Applied Functions
Section 4.5   Transformation of Functions
Section 4.6   Quadratic Functions and Their Extreme Values
Section 4.7   Using Graphing Devices to Find Extreme Values
Section 4.8   Combining Functions
Section 4.9   Composition of Functions
Section 4.10   One-to-one Functions and Their Inverses

Section 4.1 Functions

Key Ideas
A. Functions.
B. Domains.
C. Ranges.

A. A **function**, f is a rule that assigns to each element x in a set A *exactly* one element, called $f(x)$, in a set B. That is, $x \in A$ and $f(x) \in B$. Set A is called the **domain** and set B is called the **range**. '$f(x)$' is read as 'f *of* x' or 'f *at* x'. Since $f(x)$ depends on the value x, $f(x)$ is the called the dependent variable and x is called the independent variable.

1. Find the values of f defined by the equation
   $f(x) = 3x - 4$.

   a) $f(-2)$

   Substituting -2 for x
   $f(-2) = 3(-2) - 4 = -6 - 4$
   $= -10$

   b) $f(3)$

   Substituting 3 for x
   $f(3) = 3(3) - 4 = 9 - 4$
   $= 5$

   c) $f\left(\dfrac{2}{3}\right)$

   Substituting $\dfrac{2}{3}$ for x
   $f\left(\dfrac{2}{3}\right) = 3\left(\dfrac{2}{3}\right) - 4 = 2 - 4$
   $= -2$

2. Find the values of f defined by the equation
   $f(x) = \dfrac{3x^2 + x}{2 - x}$.

   a) $f(-2)$

   Substituting -2 for x
   $f(-2) = \dfrac{3(-2)^2 + (-2)}{2 - (-2)} = \dfrac{12 - 2}{2 + 2}$
   $= \dfrac{10}{4} = \dfrac{5}{2}$

   b) $f(4)$

   Substituting 4 for x
   $f(4) = \dfrac{3(4)^2 + (4)}{2 - (4)} = \dfrac{48 + 4}{2 - 4}$
   $= \dfrac{52}{-2} = -26$

   c) $f(0)$

   Substituting 0 for x
   $f(0) = \dfrac{3(0)^2 + (0)}{2 - (0)} = \dfrac{0 + 0}{2 - 0}$
   $= \dfrac{0}{2} = 0$

B. If not explicit stated, the **domain** of a function is the set of all real numbers for which the function makes sense and defines a real number. The domain is the 'input' to the function.

3. Find the domain of the function
$$f(x) = \frac{4}{x+2}.$$

The only values of x for which f does not make sense occur when the denominator is 0, that is, when $x + 2 = 0$ or $x = -2$. So the domain of f is $(-\infty, -2) \cup (-2, \infty)$. This can also be expressed as: $\{x \mid x \neq -2\}$ or $x \neq -2$.

4. Find the domain of the function
$$f(x) = \frac{x-9}{x^3 - 25x}.$$

Again, the only values of x for which f does not make sense occurs when the denominator is 0.
$x^3 - 25x = 0$
$x(x - 5)(x + 5) = 0$ Factoring;
$x = 0$ $x - 5 = 0$ $x + 5 = 0$
$x = 0$ $x = 5$ $x = -5$.
Domain is $\{x \mid x \neq 0, x \neq -5, x \neq 5\}$
or can be expressed as
$(-\infty, -5) \cup (-5, 0) \cup (0, 5) \cup (5, \infty)$

5. Find the domain of the function
$$f(x) = \sqrt{x^2 - 4}.$$

We are looking for the values of x for which the function makes sense and *defines a real number*.
So $x^2 - 4 \geq 0$ *Since the we can only find the*
$x^2 - 4 = 0$ *square root of a nonnegative number.*
$x = \pm 2$ Check the intervals.

Intervals	Test pt.	Test
$(-\infty, -2]$	-5	$(-5)^2 - 4 \stackrel{?}{\geq} 0$; yes
$[-2, 2]$	0	$(0)^2 - 4 \stackrel{?}{\geq} 0$; no
$[2, \infty)$	5	$(5)^2 - 4 \stackrel{?}{\geq} 0$; yes

Domain is $(-\infty, -2] \cup [2, \infty)$.

C. The **range** is the set of all possible values, $f(x)$, as x varies throughout the domain of f. The range is sometimes called the **image**, because it is the 'image of x under the rule f'. The range depends on the domain, so the range is referred to as the *dependent* variable. The range is the 'output' of the function.

6. Find the range of the function $f(x) = x^2 + 1$ for each domain.

   a) $[0, 4]$

   Range: $[1, 17]$.
   The small value of f occurs when $x = 0$ while the largest value of f (on this interval) occurs when $x = 4$.

   b) $[0, 4)$

   Range: $[1, 17)$.
   Notice the difference in the ranges for the small difference in the domain.

   c) $[-4, 4]$

   Range: $[1, 17]$.

   d) $[-7, 6]$

   Range: $[1, 50]$.

Section 4.2 Graphs of Functions

Key Ideas
A. Graphs of linear functions and other functions.
B. Vertical line test.
C. Increasing and decreasing functions.
D. Piecewise defined functions.
E. Symmetry, "odd", and "even" functions.

A. If f is a function with domain A, its **graph** is the set of ordered pairs, $\{(x, f(x)) \mid x \in A\}$. This means, the graph of f is all the points (x, y) where $y = f(x)$ and x is in the domain of f. A function defined by the form $f(x) = mx + b$ is called a **linear function** because the equation $y = mx + b$ is a line with slope m and y-intercept b. When $m = 0$, the linear function becomes the **constant function**, $f(x) = b$.

1. If $f(x) = 3x + 1$ find the domain and range and sketch the graph of f.

   This is a linear function with slope 4 and y-intercept 1. Generate some points and plot.

x	$f(x)$
-1	-2
0	1
1	4
2	7

   Domain: all real numbers.
   Range: all real numbers

   Note: The domain for linear functions is the reals.

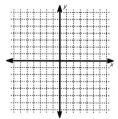

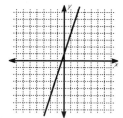

2. If $f(x) = \sqrt{2x + 1}$ find the domain and range and sketch the graph of f.

   This is not a linear function. We need to find the domain *before* we can generate points to plot.

   Domain: $2x + 1 \geq 0 \Rightarrow 2x \geq -1 \Rightarrow x \geq -\dfrac{1}{2}$

x	$f(x)$
$-\dfrac{1}{2}$	0
0	1
$\dfrac{3}{2}$	2
4	3

   Domain: $\left[-\dfrac{1}{2}, \infty\right)$
   Range: $[0, \infty)$

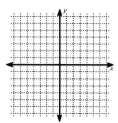

3. If $f(x) = -4$ find the domain and range and sketch the graph of f.

This is a constant function, every input value gives the same output value, in this case -4

x	$f(x)$
-100	-4
0	-4
5	-4

Domain: all real numbers.
Range: $\{-4\}$

Note: The domain for all constant functions are the reals.

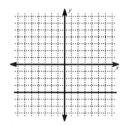

B. The **Vertical Line Test** states that 'a curve in the plane is the graph of a function if and only if no vertical line intersects the curve more than once'. It works because a function defines a unique value, $f(a)$, so if a vertical line ($x = a$) intersect a graph in more than one place it cannot be an unique value. The vertical line test is a very easy test to apply.

4. Use the vertical line test to determine which curves are graphs of functions of x.

a)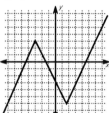

Pass vertical line test. This is the graph of a function.

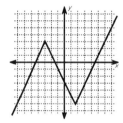

b)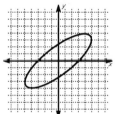

Fails vertical line test. This is not the graph of a function.

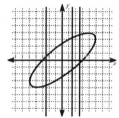

c)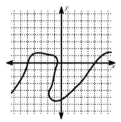

Fails vertical line test. This is not the graph of a function.

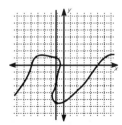

C. Graphs of functions are described by what the function does as x increases over an interval in the domain.
A function f is call **increasing** on an interval I if $f(x_1) < f(x_2)$ whenever $x_1 < x_2$ in I.
A function f is call **decreasing** on an interval I if $f(x_1) > f(x_2)$ whenever $x_1 < x_2$ in I.

5. State the interval in which the function whose graph is shown below is increasing or decreasing.

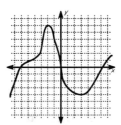

The function is *increasing* on $x < -2$ and on $3 < x$.

The function is *decreasing* on $-2 < x < 3$.

D. A function is called **piecewise** defined if its definition is defined differently for distinct intervals of its domain.

6. Sketch the graph of the function defined by:

$$f(x) = \begin{cases} x + 1 & \text{if } x < 0 \\ x^2 + 1 & \text{if } 0 \leq x \end{cases}$$

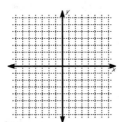
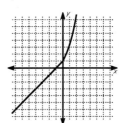

7. Sketch the graph of the function defined by:

$$f(x) = \begin{cases} \sqrt{-5 - 3x} & \text{if } x \leq -2 \\ \sqrt{5 + 2x} & \text{if } -2 < x < 2 \\ -x + 8 & \text{if } 2 \leq x \end{cases}$$

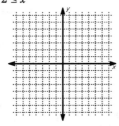
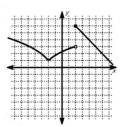

E. Symmetry often helps in graphing a function. An **even function** f is a function for which $f(-x) = f(x)$ for all x in its domain. The graph of an even function is symmetric about the y-axis. An **odd function** f is a function for which $f(-x) = -f(x)$ for all x in its domain.

The graph of an odd function is symmetric about the origin. Before graphing, quickly check to see if the function is odd, even, or neither.

8. Check each function to see if it is odd, even, or neither. Then graph.

   a) $f(x) = x^2 - 4$

   $f(x) = x^2 - 4$
   $f(-x) = (-x)^2 - 4 = x^2 - 4 = f(x)$
   $f(-x) = f(x)$, so f is even.

   b) $f(x) = \dfrac{8x}{x^2 + 1}$

   $f(x) = \dfrac{8x}{x^2 + 1}$
   $f(-x) = \dfrac{-8x}{(-x)^2 + 1} = -\dfrac{8x}{x^2 + 1} = -f(x)$
   $f(-x) = -f(x)$, so f is odd.

   c) $f(x) = |x|$

   $f(x) = |x|$
   $f(-x) = |-x| = |x| = f(x)$
   $f(-x) = f(x)$, so f is even.

   d) $f(x) = x^3 - x^2 - 8x + 4$

   $f(x) = x^3 - x^2 - 8x + 4$
   $f(-x) = (-x)^3 - (-x)^2 - 8(-x) + 4 = -x^3 - x^2 + 8x + 4$
   while $-f(x) = -x^3 + x^2 + 8x - 4$
   Since $f(-x) \neq -f(x)$ and $f(-x) \neq f(x)$, f is neither odd nor even.

Section 4.3   Graphing Calculators and Computers

Key Ideas
A. Viewing rectangles and the domain and range.
B. Graphing functions defined by more than one equation.
C. Pitfalls of graphing calculators.

A. Calculator and computer graph by plotting points. Sometimes the points are connected by straight line, other times just the points are displayed. The viewing area of a calculator or computer is referred to as the **viewing rectangle**. This is a $[a, b]$ by $[c, d]$ portion of the coordinated axis. Choosing the *appropriate* viewing rectangle is one of the most important aspects of using a graphing calculator.

1. Use a graphing calculator to draw the graph of the function $f(x) = \sqrt{x^2 + 9}$ in the following viewing rectangles.

    a) $[-3, 3]$ by $[-3, 3]$

The solutions shown below are representations of what you should see on your graphing calculator.

Only the point (0, 3) should appear.

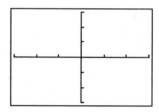

    b) $[-5, 5]$ by $[-5, 5]$

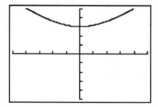

    c) $[-5, 5]$ by $[0, 10]$

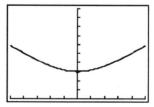

82

2. Determine an appropriate viewing rectangle for the function $f(x)=\sqrt{9-4x^2}$. Graph f in this rectangle.

First find the domain of f.
$9 - 4x^2 \geq 0 \qquad 9 - 4x^2 = 0$
$(3 - 2x)(3 + 2x) = 0$

$3 - 2x = 0 \Rightarrow 3 = 2x \Rightarrow \dfrac{3}{2} = x$

or $3 + 2x = 0 \Rightarrow 3 = -2x \Rightarrow -\dfrac{3}{2} = x$

After checking intervals, domain: $\left[-\dfrac{3}{2}, \dfrac{3}{2}\right]$.

Range; when $x = \pm\dfrac{3}{2}$, $f(x) = 0$; f is the larges when $x = 0$; so range: $[0, 3]$

Viewing rectangle $\left[-\dfrac{3}{2}, \dfrac{3}{2}\right]$ by $[0, 3]$

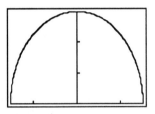

B. Piecewise-defined functions are graphed on a graphing calculator by graphing all functions and then using the intervals to sketch the graph.

3. Graph the function below on a graphing calculator. Then sketch the graph below.

$f(x) = \begin{cases} -x + 1 & \text{if } x < 0 \\ \sqrt{x^2 + 1} & \text{if } 0 \leq x \end{cases}$

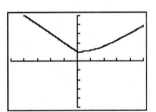

4. Graph the function below on a graphing calculator. Then sketch the graph below.

$f(x) = |x^2 - 4|$

Since $|a| = a$ or $-a$ depending on the sign of a. Graph the equations $y = x^2 - 4$ and $y = (x^2 - 4) = -x^2 + 4$. Since $f(x) \geq 0$, pick the parts where $f(x)$ is positive.
Graph of both. $f(x) \geq 0$

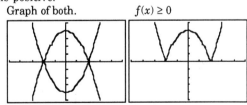

83

5. Graph the ellipse $\frac{x^2}{4} + \frac{y^2}{9} = 1$ by graphing the top half and the bottom half of the ellipse.

Start by solving the equation $\frac{x^2}{4} + \frac{y^2}{9} = 1$ for y.

$$\frac{y^2}{9} = 1 - \frac{x^2}{4} \qquad \sqrt{\frac{y^2}{9}} = \pm\sqrt{1 - \frac{x^2}{4}}$$

$$\frac{y}{3} = \pm\sqrt{1 - \frac{x^2}{4}} \qquad y = \pm 3\sqrt{1 - \frac{x^2}{4}}$$

Top half: $y = 3\sqrt{1 - \frac{x^2}{4}}$

Bottom half: $y = -3\sqrt{1 - \frac{x^2}{4}}$

[–5, 5] by [–4, 4]

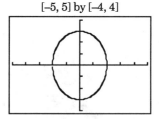

C. There are many pitfalls to using graphing calculators. One of these is a misleading impression of the graph of a function based on an inappropriate viewing rectangle. Another pitfall occurs when the function has a vertical asymptote. Vertical asymptotes occur at the point where the denominator is zero. If the viewing rectangle is to large, this will appear as a vertical part of the graph.

6. Graph the ellipse $\frac{x^2}{4} + \frac{y^2}{9} = 1$ on a graphing calculator in each of the viewing rectangles, then sketch the graph below.
   a) [2, –2] by [3, –3]

This example shows how distortion can change the view of a graph.

a)   b)

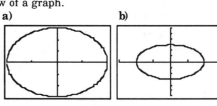

   b) [3, –3] by [6, –6]

c)

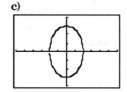

   c) [6, –6] by [4, –4]

   d) Now graph the ellipse.

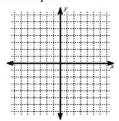

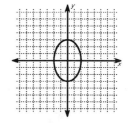

84

7. Graph the function below on a graphing calculator. Then sketch the graph below.

$$f(x) = \frac{-2x^2 - 7}{x^2 + 3x - 10}$$

f has vertical asymptotes at $x = -5$ and $x = 2$. Here f is viewed in the $[-12, 12]$ by $[-10, 10]$ rectangle.

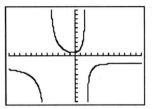

8. Graph the function below on a graphing calculator. Then sketch the graph below.

$$f(x) = \frac{4x^2 - 3}{x^2 - 6x + 9}$$

f has a vertical asymptote at $x = 3$. Both ends of the graph go up at this asymptote. Here f is viewed in the $[-4, 16]$ by $[-2, 12]$ rectangle.

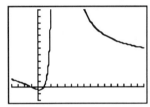

Section 4.4 Applied Functions

Key Ideas
A. Mathematical models.
B. Direct and inverse variation.

A. A **mathematical model** is a function that describes, at least approximately, the dependence of one physical quantity to another.
The following steps are helpful in setting up applied functions.
1. Understand the problem. Read the problem *until* you clearly understand it.
2. Draw a diagram or make a table. Label both the known and unknown quantities.
3. Introduction notation. Clearly define what each variable stands for; this will help set up the equation(s).

1. Express the surface area of a cube as a function of its volume.

   Let V, S, and x be the volume, surface area, and length of a side of the cube.
   Since a cube has 6 sides and each is a square, the surface area is: $S = 6x^2$
   $V = x^3$
   Solving for x, $x = \sqrt[3]{V}$
   So $S = 6\left(\sqrt[3]{V}\right)^2 = 6V^{2/3}$

2. The diagonal of a rectangle is 30 meters. Express the area of the rectangle as a function of its width w.

   Since the sides of the rectangle are perpendicular, the diagonal of the rectangle is related to the sides of the rectangle by the Pythagorean Theorem:
   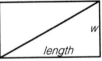
   $(length)^2 + (width)^2 = (diagonal)^2$.
   So $(length)^2 = (diagonal)^2 - (width)^2$
   Using $diagonal = 30$ and $width = w$, solve for $length$:
   $(length)^2 = 30^2 - w^2$ or $length = \sqrt{900 - w^2}$
   So the Area $= w = \sqrt{900 - w^2}$ sq. meters

B. y is **directly proportional** to x if x and y are related by the equation $y = kx$, $k \neq 0$. The constant k is called the *constant of proportionality*.

3. Federal excise tax on gasoline is 11¢ per gallon. Write an equation that directly relates the tax as a function of gallons

   Let $T =$ the tax and let $g =$ gallons of gasoline.
   Since *tax* is 11¢ per gallon
   $t = 0.11 \cdot g$
   Here 0.11 is the constant of proportionality.

4. The volume of liquid in a soda can is directly related to the height of the liquid in the can.

   a) If the can is 12.5 cm tall and contains 355 ml when full, determine the constant of proportionality and write the equation for the variation.

   Let h = height, V = volume.
   $V = kh$
   When $V = 355$, $h = 12.5$. So $355 = 12.5k$
   and $k = 28.4$
   $V = 28.4h$

   b) How much is left in the can when the liquid is 3 cm deep?

   Setting $h = 3$ in the equation yields:
   $V = 28.4 \cdot 3$
   $= 85.2$ ml

   c) How deep is the liquid when there is 300 ml left?

   Setting $V = 300$ and solving for h yields:
   $300 = 28.4h$
   So $h = 10.56$ cm

B. y is **inversely proportional** to x if x and y are related by the equation $y = \dfrac{k}{x}$ where $k \neq 0$.

5. Write an equation that relates the width to the length is all rectangle whose area is 4 square units.

   Let w = width and let l length.
   Since $A = wl$ we have $4 = wl$ and
   $w = \dfrac{4}{l}$.

Section 4.5 Transformation of Functions

Key Ideas
A. Vertical shifts of graphs.
B. Horizontal shifts of graphs.
C. Vertical stretching, shrinking, and reflecting.

A. When $c > 0$, $y = f(x) + c$ moves the graph of $y = f(x)$ vertically c units up, and $y = f(x) - c$ moves the graph of $y = f(x)$ vertically c units down.

1. Graph the functions $f(x) = x^3$ and $g(x) = x^3 + 4$ on the same coordinate axis.

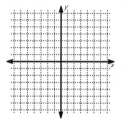

2. Graph the functions $f(x) = |x|$ and $g(x) = |x| - 2$ on the same coordinate axis.

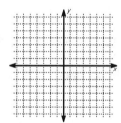
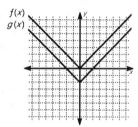

B. When $c > 0$, $y = f(x - c)$ moves the graph of $y = f(x)$ horizontally c units to the right, and $y = f(x + c)$ moves the graph of $y = f(x)$ horizontally c units to the left.

3. Let $f(x) = x^2$. Express the function $f(x - 3)$ and $f(x + 2)$. Sketch all three on the same coordinate axis.

$f(x - 3) = (x - 3)^2 = x^2 - 6x + 9$
$f(x + 2) = (x + 2)^2 = x^2 + 4x + 4$

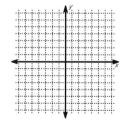
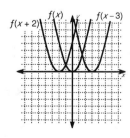

4. Let $f(x) = \dfrac{5}{x^2 + 1}$. Express the function $f(x-2)$ and $f(x+1)$. Sketch all three on the same coordinate axis.

$f(x-2) = \dfrac{5}{(x-2)^2 + 1} = \dfrac{5}{x^2 - 4x + 5}$

$f(x+1) = \dfrac{5}{(x+1)^2 + 1} = \dfrac{5}{x^2 + 2x + 2}$

Note: Both $f(x-2)$ and $f(x+1)$ are much easier to graph using the horizontal shifts.

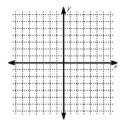
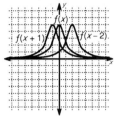

C. For $c > 1$, $y = cf(x)$ stretches the graph of $y = f(x)$ vertically by a factor of c. $y = \dfrac{1}{c} f(x)$ shrinks the graph of $y = f(x)$ vertically by a factor of c. And $y = -f(x)$ reflects the graph of $y = f(x)$ over the x-axis.

5. Let $f(x) = 2x - 1$. Find $3f(x), \dfrac{1}{2}f(x)$. Sketch all three graphs on the same coordinate axis.

$3f(x) = 3(2x - 1) = 6x - 3$

$\dfrac{1}{2}f(x) = \dfrac{1}{2}(2x - 1) = x - \dfrac{1}{2}$

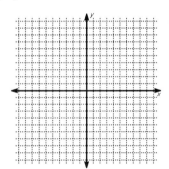
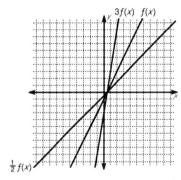

6. Let $f(x) = \sqrt{x} - 1$. Find $2f(x), -2f(x)$. Sketch all three graphs on the same coordinate axis.

$2f(x) = 2(\sqrt{x} - 1) = 2\sqrt{x} - 2$

$-2f(x) = -2(\sqrt{x} - 1) = -2\sqrt{x} + 2$

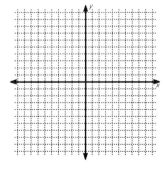
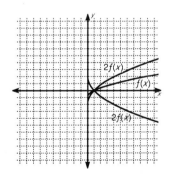

7. Let $f(x) = 3x + \dfrac{2}{x}$. Find $-f(x)$, $-\dfrac{1}{2}f(x)$. Sketch all three graphs on the same coordinate axis.

$-f(x) = -\left(3x + \dfrac{2}{x}\right) = -3x - \dfrac{2}{x}$

$-\dfrac{1}{2}f(x) = -\dfrac{1}{2}\left(3x + \dfrac{2}{x}\right) = -\dfrac{3}{2}x - \dfrac{1}{x}$

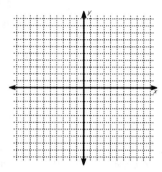
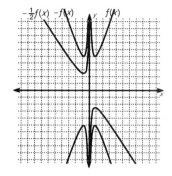

8. Let $f(x) = \dfrac{5}{x^2 + 1}$. Express the function $\dfrac{1}{2}f(x)$ and $-2f(x)$. Sketch all three on the same coordinate axis.

$\dfrac{1}{2}f(x) = \dfrac{5}{2(x^2 + 1)}$

$-2f(x) = \dfrac{-10}{x^2 + 1}$

Compare these with the transformations in problem 4.

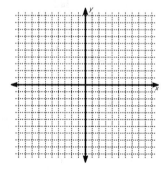
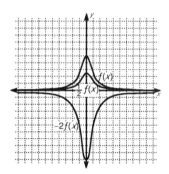

Section 4.6   Quadratic Functions and Their Extreme Values

Key Ideas
A. Quadratic functions.
B. Maximum and minimum values.

A. A **quadratic function** is a function of the form $f(x) = ax^2 + bx + c$, where a, b and c are real numbers, and $a \neq 0$. It is often easier to graph and work with quadratic functions when we have completed the square and expressed f in the form $f(x) = a(x - h)^2 + k$.

1. Express f in the form $f(x) = a(x - h)^2 + k$ and sketch the graph of f.

   a) $f(x) = x^2 + 6x$

   $$f(x) = x^2 + 6x$$
   $$= (x^2 + 6x + 9) - 9$$
   $$= (x + 3)^2 - 9$$

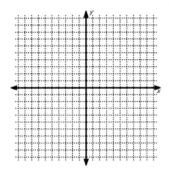
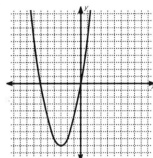

   b) $f(x) = x^2 - 4x - 2$

   $$f(x) = x^2 - 4x - 2$$
   $$= (x^2 - 4x + 4) - 2 - 4$$
   $$= (x - 2)^2 - 6$$

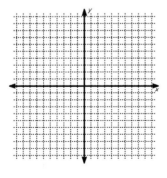
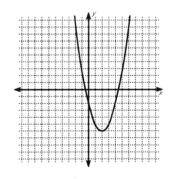

c) $f(x) = 2x^2 - 6x - \frac{5}{2}$

$f(x) = 2x^2 - 6x - \frac{5}{2} = 2(x^2 - 3x) - \frac{5}{2}$
$= 2\left(x^2 - 3x + \frac{9}{4}\right) - \frac{5}{2} - \frac{9}{2}$
$= 2\left(x - \frac{3}{2}\right)^2 - 7$

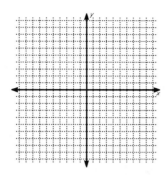
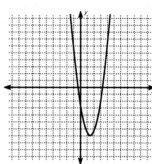

B. When $a > 0$, the parabola (quadratic function), $f(x) = a(x - h)^2 + k$, opens up. The lowest point is the vertex (h, k), so the **minimum value** of the function occurs when $x = h$, and this minimum value is $f(h) = k$. Also $f(x) = a(x - h)^2 + k \geq k$ for all x.

When $a < 0$, the parabola (quadratic function), $f(x) = a(x - h)^2 + k$, opens down. The highest point is the vertex (h, k), so the **maximum value** of the function occurs when $x = h$ and this maximum value is $f(h) = k$. Also $f(x) = a(x - h)^2 + k \leq k$ for all x.

2. Determine the maximum or minimum value of f by completing the square. Find the intercepts and then sketch the graph.

   a) $f(x) = x^2 + 5x + 4$

$f(x) = x^2 + 5x + 4 = (x^2 + 5x) + 4$
$= \left(x^2 + 5x + \frac{25}{4}\right) + 4 - \frac{25}{4} = \left(x + \frac{5}{2}\right)^2 - \frac{9}{4}$

Minimum value: $-\frac{9}{4}$ at $x = -\frac{5}{2}$.

To find the x-intercepts set $f = 0$ and solve.
$f(x) = x^2 + 5x + 4 = 0$
$(x + 1)(x + 4) = 0$
$x + 1 = 0 \quad x + 4 = 0$
$x = -1 \quad\quad x = -4$
x-intercepts: -1 and -4
To find the y-intercept evaluate $f(0)$.
y-intercepts: $f(0) = 4$

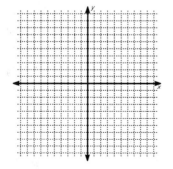
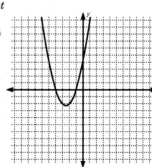

b) $f(x) = 3x^2 - 6x + 4$

$$f(x) = 3x^2 - 6x + 4 = 3(x^2 - 2x) + 4$$
$$= 3(x^2 - 2x + 1) + 4 - 3$$
$$= 3(x - 1)^2 + 1$$

Minimum value: 1 at $x = 1$.

Since $f(x) = 3(x - 1)^2 + 1$, and since the sum of two squares does *not* factor, there are **no** x-intercepts.

y-intercepts: 4

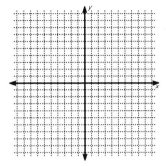

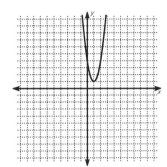

c) $f(x) = -2x^2 + 5x$

$$f(x) = -2x^2 + 5x = -2\left(x^2 - \frac{5}{2}x\right)$$
$$= -2\left(x^2 - \frac{5}{2}x + \frac{25}{16}\right) + \frac{25}{8}$$
$$= -2\left(x - \frac{5}{4}\right)^2 + \frac{25}{8}$$

Maximum value: $\frac{25}{8}$ at $x = \frac{5}{4}$.

$f(x) = -2x^2 + 5x = 0$
$x(-2x + 5) = 0$
$x = 0 \quad -2x + 5 = 0$
$ -2x = -5$
$ x = \frac{5}{2}$

x-intercepts: 0 and $\frac{5}{2}$ *Don't forget the zero!*

y-intercepts: 0

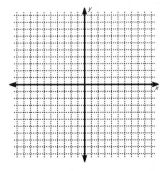

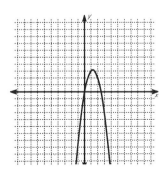

93

d) $f(x) = -5x^2 - 10x + 1$

$f(x) = -5x^2 - 10x + 1 = -5(x^2 + 2x) + 1$
$= -5(x^2 + 2x + 1) + 1 + 5$
$= -5(x + 1)^2 + 6$

Maximum value: 6 at $x = -1$.

$f(x) = -5x^2 - 10x + 1 = 0$

Since this is hard to factor, use the quadratic formula to find the -intercepts.

$x = \dfrac{-(-10) \pm \sqrt{(-10)^2 - 4(-5)(1)}}{2(-5)}$

$x = \dfrac{10 \pm \sqrt{100 + 20}}{-10} = \dfrac{10 \pm \sqrt{120}}{-10}$

$= -1 \pm \dfrac{2\sqrt{30}}{5} \approx -1 \pm 2.19$

x-intercepts: -3.19 and 1.19
y-intercepts: $f(0) = 1$

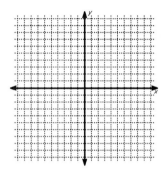

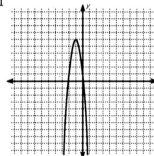

3. A farmer wants to enclose a two rectangular field on three sides by a fence and divided it into smaller rectangular fields by a fence perpendicular to the fourth side. He has 3000 yards of fencing. Find the dimension of the field so that the total enclosed area is a maximum.

First make a drawing and label the sides.

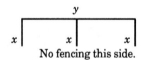

No fencing this side.

Fencing $3x + y = 3000$
Total area, $A = xy$
Solve the first for y and substitute into the second yields:
$y = 3000 - 3x$
$A = x(3000 - 3x) = -3x^2 + 3000x$

Since the leading coefficient is negative, A will have a maximum. Now complete the square:

$A = -3(x^2 - 1000x)$
$= -3(x^2 - 1000x + 250000) + 750000$
$= -3(x - 500)^2 + 750000$

So A is a maximum of 750000 sq. yds when $x = 500$ yds and $y = 3000 - 3(500) = 1500$ yds.

Section 4.7 Using Graphing Devices to Find Extreme Values

Key Ideas
A. Local extreme values.

A. A point a is called a **local maximum** of the function f if $f(a) > f(x)$ for all x in some interval containing a. Likewise, a **local minimum** of the function f is a point a where $f(a) < f(x)$ for all x in some interval containing a. When using a graphing device, these local extreme values can be found by looking for the highest (or lowest) point *within* the viewing rectangle.

1. Find the approximate local maximum and minimum values of $f(x) = x^3 - 2x + 4$.

   Local minimum:
   $x = \dfrac{\sqrt{6}}{3} \approx 0.816$
   $y = -\dfrac{4}{9}\sqrt{6} + 4 \approx 2.911$

   [0,1] by [2.5, 3]

   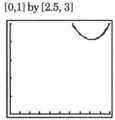

   Local maximum:
   $x = -\dfrac{\sqrt{6}}{3} \approx -0.816$
   $y = \dfrac{4}{9}\sqrt{6} + 4 \approx 5.089$

   [−1, −0.5] by [5, 5.5]

   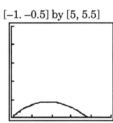

2. Find the approximate local maximum and minimum values of $f(x) = x^4 - 4x^3 + 16x$.

   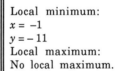
   Local minimum:
   $x = -1$
   $y = -11$
   Local maximum:
   No local maximum.

   [−2, 0] by [−12, −10]

   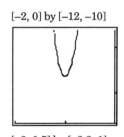

3. Find the approximate local maximum and minimum values of $f(x) = x^3 + 3x^2 - 2\sqrt{x^2 + 1}$.

   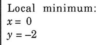
   Local minimum:
   $x = 0$
   $y = -2$

   [−2, 0.5] by [−2.2, 1]

   Local maximum:
   $x \approx -1.66$
   $y \approx -0.18$

   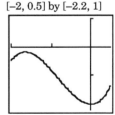

Section 4.8   Combining Functions

Key Ideas
A. Addition of functions.
B. Subtraction of functions.
C. Multiplication of functions.
D. Division of functions.

A. Let $f(x)$ be a function with domain A and let $g(x)$ be a function with domain B. Then the **sum** of the functions f and g is $(f+g)(x) = f(x) + g(x)$ and has domain $A \cap B$.

1. Let $f(x) = 8x + 3$ and $g(x) = x^2 - 4$. Find $f + g$ and find its domain.

    $(f + g)(x) = f(x) + g(x)$
    $= (8x + 3) + (x^2 - 4)$
    $= x^2 + 8x - 1$
    Domain: Since the domain for both f and g is the real numbers, the domain of $f + g$ is the real numbers.

2. Let $f(x) = -(x - 2)^2$ and $g(x) = |x|$. Find $f + g$ and find its domain.

    $(f + g)(x) = f(x) + g(x)$
    $= -(x - 2)^2 + |x|$
    Domain: Since the domain for both f and g is the real numbers, the domain of $f + g$ is the real numbers.

3. Let $f(x) = \sqrt{x + 3}$ and $g(x) = \sqrt{9 - x^2}$. Find $f + g$ and find its domain.

    $(f + g)(x) = f(x) + g(x)$
    $= \left(\sqrt{x + 3}\right) + \left(\sqrt{9 - x^2}\right)$
    $= \sqrt{x + 3} + \sqrt{9 - x^2}$
    Domain of f: $x + 3 \geq 0$ so $x \geq -3$ or $[-3, \infty]$
    Domain of g: $9 - x^2 \geq 0$ After checking intervals, $[-3, 3]$
    Domain of $f + g$: $[-3, 3]$

B. Let $f(x)$ be a function with domain A and let $g(x)$ be a function with domain B. Then the **difference** of the functions f and g is $(f - g)(x) = f(x) - g(x)$ and has domain $A \cap B$.

4. Let $f(x) = \sqrt{x^2 - 1}$ and $g(x) = \sqrt{16 - x^2}$. Find $f - g$ and find its domain.

$(f - g)(x) = f(x) - g(x)$
$= \left(\sqrt{x^2 - 1}\right) - \left(\sqrt{16 - x^2}\right)$
$= \sqrt{x^2 - 1} - \sqrt{16 - x^2}$
Domain of f: $x^2 - 1 \geq 0$ After checking intervals, $[-\infty, -1] \cup [1, \infty]$
Domain of g: $16 - x^2 \geq 0$ After checking intervals, $[-4, 4]$
Domain of $f - g$: $[-4, -1] \cup [1, 4]$

5. Let $f(x) = \dfrac{2}{x - 3}$ and $g(x) = \dfrac{1}{x + 5}$. Find $f - g$ and find its domain.

$(f - g)(x) = f(x) - g(x)$
$= \left(\dfrac{2}{x - 3}\right) - \left(\dfrac{1}{x + 5}\right)$
$= \dfrac{2}{x - 3} - \dfrac{1}{x + 5}$
Domain of f: $x - 3 \neq 0$ or $x \neq 3$
Domain of g: $x + 5 \neq 0$ or $x \neq -5$
Domain of $f - g$: $\{x \mid x \neq 3, x \neq -5\}$

6. Let $f(x) = x$ and $g(x) = \dfrac{1}{x^2 - 4}$. Find $f - g$ and find its domain.

$(f - g)(x) = f(x) - g(x)$
$= (x) - \left(\dfrac{1}{x^2 - 4}\right)$
$= x - \dfrac{1}{x^2 - 4}$
Domain of f: All real numbers
Domain of g: $x^2 - 4 \neq 0$ $\{x \mid x \neq \pm 2\}$
Domain of $f - g$: $\{x \mid x \neq \pm 2\}$

C. Let $f(x)$ be a function with domain A and let $g(x)$ be a function with domain B. Then the **product** of the functions f and g is $(fg)(x) = f(x)g(x)$ with domain $A \cap B$.

7. Let $f(x) = x + 2$ and $g(x) = \dfrac{2}{x + 1} - 4$. Find fg and find its domain.

$(fg)(x) = f(x)g(x)$
$= (x + 2)\left(\dfrac{2}{x + 1} - 4\right)$
$= \dfrac{2x + 4}{x + 1} - 4x + 8$
Domain of f: All real numbers
Domain of g: $x + 1 \neq 0$ or $x \neq -1$
Domain of fg: $\{x \mid x \neq -1\}$

8. Let $f(x) = \sqrt{x+12}$ and $g(x) = \sqrt{x-5}$. Find fg and find its domain.

$(fg)(x) = f(x)g(x)$
$= (\sqrt{x+12})(\sqrt{x-5})$
$= \sqrt{x^2 + 7x - 60}$
Domain of f: $x + 12 \geq 0$ so $x \geq -12$
Domain of g: $x - 5 \geq 0$ so $x \geq 5$
Domain of fg: $x \geq 5$ or $[5, \infty]$

D. Let $f(x)$ be a function with domain A and let $g(x)$ be a function with domain B. Then the **quotient** of the functions f and g is $\left(\dfrac{f}{g}\right)(x) = \dfrac{f(x)}{g(x)}$ with domain $\{x \in A \cap B \mid g(x) \neq 0\}$

9. Let $f(x) = \sqrt{x-1}$ and $g(x) = x^2 - 4$.

   a) Find $\dfrac{f}{g}$ and find its domain.

   $\dfrac{f(x)}{g(x)} = \dfrac{\sqrt{x-1}}{x^2 - 4}$
   Domain:
   Numerator: $x - 1 \geq 0$, $x \geq 1$.
   Denominator: $x^2 - 4 \neq 0$
   $x^2 - 4 = 0$
   $(x-2)(x+2) = 0$
   $x = 2$ or $x = -2$ so $x \neq \pm 2$
   So domain is $[1, \infty) \cap \{x \mid x \neq \pm 2\}$ or $[1, 2) \cup (2, \infty)$

   b) Find $\dfrac{g}{f}$ and find its domain.

   $\dfrac{g(x)}{f(x)} = \dfrac{x^2 - 4}{\sqrt{x-1}}$
   Domain:
   Numerator: all real numbers
   Denominator: $x - 1 > 0$, $x > 1$.
   Domain is $x > 1$ or $(1, \infty)$

10. Let $f(x) = x^2 - 9$ and $g(x) = x^2 + 9$.

    a) Find $\dfrac{f}{g}$ and find its domain.

    $\dfrac{f(x)}{g(x)} = \dfrac{x^2 - 9}{x^2 + 9}$
    Domain:
    Numerator: all real numbers.
    Denominator: all real numbers.
    Domain is all real numbers.

    Note: $x^2 + 9$, the sum of two squares, is never 0.

    b) Find $\dfrac{g}{f}$ and find its domain.

    $\dfrac{g(x)}{f(x)} = \dfrac{x^2 + 9}{x^2 - 9}$
    Domain:
    Numerator: all real numbers.
    Denominator: $x^2 - 9 \neq 0$, $x \neq \pm 3$.
    Domain is $\{x \mid x \neq \pm 3\}$

Section 4.9   Composition of Functions

Key Ideas
A. Composition of functions.

A. Given two functions f and g, the **composition function** $f \circ g$ is defined by $(f \circ g)(x) = f(g(x))$. We sometimes say that $f \circ g$ is 'f acting on $g(x)$', in other words, x is the input to g and $g(x)$ is the input to f. The domain of $f \circ g$ is the subset of the domain of g where $g(x)$ is in the domain of f. Another way of saying this, is that the domain of $f \circ g$ is those points where both $g(x)$ and $f(g(x))$ are both defined.

1. Let $f(x) = \sqrt{x - 10}$ and let $g(x) = x + 9$.
   a) Find $f \circ g$ and find its domain and range.

   $(f \circ g)(x) = f(g(x)) = f(x + 9)$
   $= \sqrt{(x + 9) - 10}$
   $= \sqrt{x - 1}$
   Domain: $x - 1 \geq 0$ so $x \geq 1$.
   Range: $f \circ g \geq 0$

   b) Find $g \circ f$ and find its domain and range.

   $(g \circ f)(x) = g(f(x)) = g(\sqrt{x-10})$
   $= \sqrt{x - 10} + 9$
   Domain: $x - 10 \geq 0$ so $x \geq 10$.
   Range: $g \circ f \geq 9$
   Note: $f \circ g \neq g \circ f$.

2. Let $f(x) = \dfrac{1}{x} - 4$ and let $g(x) = \dfrac{1}{x + 4}$.
   a) Find $f \circ g$ and find its domain and range.

   $(f \circ g)(x) = f(g(x)) = f\left(\dfrac{1}{x + 4}\right)$
   $= \dfrac{1}{\left(\dfrac{1}{x + 4}\right)} - 4$
   $= x + 4 - 4 = x$
   Domain: $x \neq -4$ (Note: $\dfrac{1}{x + 4}$ is never 0.)
   Range: $f \circ g \neq -4$

   b) Find $g \circ f$ and find its domain and range.

   $(g \circ f)(x) = g(f(x)) = g\left(\dfrac{1}{x} - 4\right)$
   $= \dfrac{1}{\left(\dfrac{1}{x} - 4\right) + 4} = \dfrac{1}{\dfrac{1}{x}} = x$
   Domain: $x \neq 0$.
   Range: $g \circ f \neq 0$

99

3. Let $f(x) = \sqrt{x^2 + x + 1}$ and let $g(x) = x^2 - 1$.

a) Find $f \circ g$ and find its domain and range.

$$(f \circ g)(x) = f(g(x)) = f(x^2 - 1)$$
$$= \sqrt{(x^2 - 1)^2 + (x^2 - 1) + 1}$$
$$= \sqrt{x^4 - x^2 + 1}$$

Domain: Since the domain of g is all real numbers and since $x^2 + x + 1 = \left(x + \frac{1}{2}\right)^2 + \frac{3}{4}$ is always positive; domain is all real numbers.

Range: $f \circ g \geq \frac{\sqrt{3}}{2}$

b) Find $g \circ f$ and find its domain and range.

$$(g \circ f)(x) = g(f(x)) = g\left(\sqrt{x^2 + x + 1}\right)$$
$$= \left(\sqrt{x^2 + x + 1}\right)^2 - 1$$
$$= (x^2 + x + 1) - 1 \quad \text{See below.}$$
$$= x^2 + x$$

Domain: From above, $x^2 + x + 1 > 0$, so the domain is all real numbers.

Range: $g \circ f \geq -\frac{1}{4}$.

Note: Since $x^2 + x + 1 > 0$,
$$\left(\sqrt{x^2 + x + 1}\right)^2 = x^2 + x + 1.$$

Section 4.10 One-to-One Functions and Their Inverses

Key Ideas
A. One-to-one functions and the horizontal line test.
B. Domain and ranges of inverse functions.
C. Finding the inverse function of a one-to-one function.
D. Graphing the inverse function of a one-to-one function.

A. A function with domain a is called a **one-to-one function** if no two elements of A have the same image; that is, $f(x_1) \neq f(x_2)$ whenever $x_1 \neq x_2$. Another way of stating this is, "If $f(x_1) = f(x_2)$, then $x_1 = x_2$". One test for determining when a function is one-to-one is called the **vertical line test** which states that a function is one-to-one if and only if *no* horizontal line intersects it graph more than once. Care must be taken to avoid some common errors in algebraically showing a function is one-to-one. While it is true that $a \neq b$ implies that $a + c \neq b + c$, $a \neq b$ and $c \neq d$ does not imply that $a + c \neq b + d$.

1. Show that $f(x) = x^2 + 2x - 3$ is *not* one-to-one by finding two points $x_1 \neq x_2$ where $f(x_1) = f(x_2)$.

   There are many answers possible. An easy pair of points can be found by setting $f(x) = 0$. Factoring;
   $x^2 + 2x - 3 = (x - 1)(x + 3) = 0$
   $x - 1 = 0$ or $x + 3 = 0$
   $x = 1$ $x = -3$
   So, $f(1) = f(-3) = 0$ and f is not one-to-one.
   Since $f(x) = (x + 1)^2 - 4$ (*complete the square*) all solutions to $f(x_1) = f(x_2)$ are of the form $|x_1 + 1| = |x_2 + 1|$.
   Example: $|(1) + 1| = |(-3) + 1| = 2$.

2. Use the horizontal line test to determine if each function is one-to-one.
   a)

   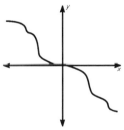

   This graph passes the horizontal line test. This is a one-to-one function.

   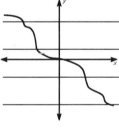

   b)

   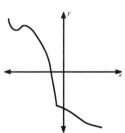

   This graph does *not* pass the horizontal line test. This function is *not* one-to-one.

   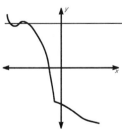

101

c)

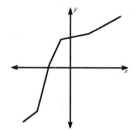

This graph passes the horizontal line test. This is a one-to-one function.

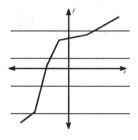

B. Let f be a one-to-one function with domain A and range B. Then its **inverse function** f^{-1} has domain B and range A and is defined by
$$f^{-1}(y) = x \quad \Leftrightarrow \quad f(x) = y \quad \text{for any } y \text{ in } B.$$
It is a common mistake to think of the -1 in f^{-1} as an exponent, $f^{-1}(x) \neq \dfrac{1}{f(x)}$.

3. Suppose f is one-to-one and $f(0) = 3$, $f(1) = 4$, $f(3) = 5$, and $f(4) = 6$.
   a) Find $f^{-1}(3)$.

   $f^{-1}(3) = 0$ since $f(0) = 3$.

   b) Find $f^{-1}(4)$.

   $f^{-1}(4) = 1$ since $f(1) = 4$.

C. If f is one-to-one function, then we find the inverse function by the following procedure.
1. Write $y = f(x)$.
2. Solve this equation for x in terms of y (if possible).
3. Interchange x and y. The resulting function is $y = f^{-1}(x)$.

4. Find the inverse function of $f(x) = \sqrt{2x - 1}$. What is the domain and range of f^{-1}?

$y = \sqrt{2x - 1}$
$y^2 = 2x - 1$ Square both sides.
$y^2 + 1 = 2x$
$\dfrac{y^2 + 1}{2} = x$

$x = \dfrac{y^2 + 1}{2} = \dfrac{1}{2}y^2 + \dfrac{1}{2}$

$y = \dfrac{1}{2}x^2 + \dfrac{1}{2}$ Interchange x and y.

$f^{-1}(x) = \dfrac{1}{2}x^2 + \dfrac{1}{2}$

Domain of f^{-1} (= range of f): $[0, \infty)$.
Range of f^{-1}: $\left[\dfrac{1}{2}, \infty\right)$

Note: Here it is easier to find the range of f instead of the domain of f^{-1}.

5. Find the inverse function of $f(x) = \dfrac{3x-1}{2+x}$.
   What is the domain and range of f^{-1}?

$y = \dfrac{3x-1}{2+x}$

$y(2+x) = 3x - 1$

$2y + xy = 3x - 1$ *Gather the terms*

$xy - 3x = -2y - 1$ *with x to one side.*

$x(y - 3) = -(2y + 1)$ *Factor out x.*

$x = -\dfrac{2y+1}{y-3}$ *Interchange x and y.*

$y = -\dfrac{2x+1}{x-3}$

$f^{-1}(x) = -\dfrac{2x+1}{x-3}$

Domain of f^{-1}: $x \neq 3$.
Range of f^{-1} (= domain of f): $f^{-1} \neq -2$
Note: Here it is easier to find the domain of f instead of the range of f^{-1}.

D. Even though we might not be able to find the formula for f^{-1}, we can still find the graph of f^{-1}. The graph of f^{-1} is obtained by reflecting the graph of f over the line $y = x$. The important thing here is that (a, b) is a point on the graph of f if and only if (b, a) is a point on the graph of f^{-1}.

6. Find the graph of f^{-1} for each graph of f shown below.

   a)

   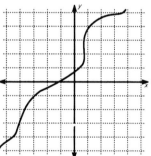
   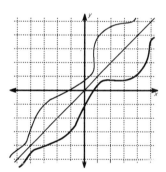

   b)

   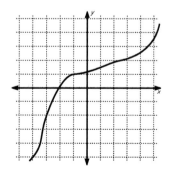
   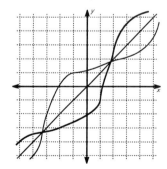

Chapter 5

Polynomials and Rational Functions

Section 5.1 Dividing Polynomials
Section 5.2 Rational Roots
Section 5.3 Using Graphing Devices to Solve Polynomial Equations
Section 5.4 Irrational Roots
Section 5.5 Complex Roots and the Fundamental Theorem of Algebra
Section 5.6 Graphing Polynomials
Section 5.7 Using Graphing Devices to Graph Polynomials
Section 5.8 Rational Functions
Section 5.9 Using Graphing Devices to Graph Rational Functions

Section 5.1   Dividing Polynomials

Key Ideas
A. Polynomial and rational functions.
B. Division algorithm.
C. The remainder theorem.
D. Factors and zeros of polynomials.
E. Synthetic division.

A. A polynomial, P, of degree n is a function of the form
$P(x) = a_n x^n + a_{n-1} x^{n-1} + \cdots + a_1 x^1 + a_0$ where $a_n, a_{n-1}, a_{n-2}, \ldots, a_1, a_0$ are constants and $a_n \ne 0$. The numbers $a_n, a_{n-1}, a_{n-2}, \ldots, a_1, a_0$ are called the **coefficients** of the polynomial. The **constant coefficient**, a_0, and the **leading coefficient**, a_n, are important in the factoring of polynomials and finding zeros of polynomials. The constant coefficient is sometimes called the **constant term**. A rational function is a function r of the form $r(x) = \dfrac{P(x)}{Q(x)}$, where P and Q are each polynomials.

1. For each polynomial determine the degree, the leading coefficient, and the constant coefficient.

   a) $P(x) = 8x^4 + 3x^3 - x^2 + 5x - 7$

   degree: 4
   leading coefficient: 8
   constant coefficient: −7

   b) $P(x) = x^7 - 3x^4 - 3x$

   degree: 7
   leading coefficient: 1
   constant coefficient: 0   Note: Even though no constant coefficient is shown, it is understood to be zero.

   c) $P(x) = 6 - 5x - 7x^3 + 2x^4 - 4x^6$

   First write in descending order
   $P(x) = -4x^6 + 2x^4 - 7x^3 - 5x + 6$
   degree: 6
   leading coefficient: −4
   constant coefficient: 6

B. Long division for polynomials is a process similar to division for numbers. The terminology of division is as follows. The **dividend** is the quantity that the **divisor** is divided into. The result is called the **quotient** and the what is left over is the **remainder**.

A very useful algorithm is the **Division Algorithm**. If $P(x)$ and $D(x)$ are polynomials, with $D(x) \neq 0$, then there exist unique polynomials $Q(x)$ and $R(x)$ such that: $P(x) = D(x) \cdot Q(x) + R(x)$, where $R(x) = 0$ or degree of $R(x)$ is less than the degree of $D(x)$. $P(x)$ is the **dividend**, $D(x)$ is the **divisor**, $Q(x)$ is the **quotient**, and $R(x)$ is the **remainder**. The important thing to remember when dividing is to insert $0x^k$ for missing terms.

2. Let $P(x) = x^4 + 4x^3 + 6x^2 - x + 6$ and $D(x) = x^2 + 2x$. Find the polynomials $Q(x)$ and $R(x)$ such that $P(x) = D(x) \cdot Q(x) + R(x)$.

$$\begin{array}{r} x^2 + 2x + 2 \\ x^2 + 2x \overline{) x^4 + 4x^3 + 6x^2 - x + 6} \\ \underline{x^4 + 2x^3} \\ 2x^3 + 6x^2 \\ \underline{2x^3 + 4x^2} \\ 2x^2 - x \\ \underline{2x^2 + 4x} \\ -5x + 6 \end{array}$$

So $x^4 + 4x^3 + 6x^2 - x + 6 =$
$(x^2 + 2x)(x^2 + 2x + 2) + (-5x + 6)$

3. Let $P(x) = x^4 + x^3 + 4$ and $D(x) = x^2 - 3$. Find the polynomials $Q(x)$ and $R(x)$ such that $P(x) = D(x) \cdot Q(x) + R(x)$.

$$\begin{array}{r} x^2 + x + 3 \\ x^2 - 3 \overline{) x^4 + x^3 + 0x^2 + 0x + 4} \\ \underline{x^4 - 3x^2} \\ x^3 + 3x^2 + 0x \\ \underline{x^3 - 3x} \\ 3x^2 + 3x + 4 \\ \underline{3x^2 - 9} \\ 3x + 13 \end{array}$$

So
$x^4 + x^3 + 4 = (x^2 - 3)(x^2 + x + 3) + (3x + 13)$

C. When $D(x) = x - c$, the Division Algorithm becomes the **Remainder Theorem**: $P(x) = (x - c) \cdot Q(x) + r$ where $r = P(c)$. This means that the remainder of $P(x)$ when divided by $(x - c)$ is the same as the functional value at $x = c$.

4. Verify the Remainder Theorem by dividing $P(x) = x^3 + 2x^2 - 10x$ by $x - 3$. Then find $P(3)$.

$$\begin{array}{r} x^2 + 5x + 5 \\ x - 3 \overline{\smash{)} x^3 + 2x^2 - 10x + 0} \\ \underline{x^3 - 3x^2 } \\ 5x^2 - 10x \\ \underline{5x^2 - 15x } \\ 5x + 0 \\ \underline{5x - 15} \\ 15 \end{array}$$

$P(3) = (3)^3 + 2(3)^2 - 10(3)$
$ = 27 + 18 - 30$
$ = 15$

5. Verify the Remainder Theorem by dividing $P(x) = x^3 + 2x^2 - 10x$ by $x + 2$. Then find $P(-2)$.

$$\begin{array}{r} x^2 - 10 \\ x + 2 \overline{\smash{)} x^3 + 2x^2 - 10x + 0} \\ \underline{x^3 + 2x^2 } \\ -10x + 0 \\ \underline{-10x - 20} \\ 20 \end{array}$$

$P(-2) = (-2)^3 + 2(-2)^2 - 10(-2)$
$ = -8 + 8 + 20$
$ = 20$

D. In the **Division Algorithm**, if $R(x) = 0$, then $P(x) = D(x) \cdot Q(x)$. $D(x)$ and $Q(x)$ are called **factors** of $P(x)$. $P(c) = 0$ if and only if $D(x) = x - c$ is a factor of $P(x)$.

6. Let $P(x) = 6x^3 - x^2 - 31x - 24$. Show that $P(-1) = 0$ and use this fact to factor $P(x)$ completely.

$P(-1) = 6(-1)^3 - (-1)^2 - 31(-1) - 24$
$= -6 - 1 + 31 - 24 = 0$
So $x = -1$ is a solution and $x + 1$ is a factor.

$$\begin{array}{r} 6x^2 - 7x - 24 \\ x+1 \overline{\smash{)}6x^3 - x^2 - 31x - 24} \\ \underline{6x^3 + 6x^2} \\ -7x^2 - 31x \\ \underline{-7x^2 - 7x} \\ -24x - 24 \\ \underline{-24x - 24} \\ 0 \end{array}$$

$P(x) = (x + 1)(6x^2 - 7x - 24)$
$= (x + 1)(3x - 8)(2x + 3)$

7. Let $P(x) = 2x^3 - 7x^2 - 7x + 30$. Show that $P(-2) = 0$ and use this fact to factor $P(x)$ completely.

$P(-2) = 2(-2)^3 - 7(-2)^2 - 7(-2) + 30$
$= -16 - 28 + 14 + 30 = 0$
So $x = -2$ is a solution and $x + 2$ is a factor.

$$\begin{array}{r} 2x^2 - 11x + 15 \\ x+2 \overline{\smash{)}2x^3 - 7x^2 - 7x + 30} \\ \underline{2x^3 + 4x^2} \\ -11x^2 - 7x \\ \underline{-11x^2 - 22x} \\ 15x + 30 \\ \underline{15x + 30} \\ 0 \end{array}$$

$P(x) = (x + 2)(2x^2 - 11x + 15)$
$= (x + 1)(2x - 5)(x - 3)$

8. Find a polynomial of degree 5 that has zeros at $-5, -2, 1, 2,$ and 3.

Since $P(c) = 0$ if and only if $x - c$ is a factor;
$P(x) = [x - (-5)][x - (-2)][x - 1][x - 2][x - 3]$
$= (x + 5)(x + 2)(x - 1)(x - 2)(x - 3)$
$= x^5 + x^4 - 21x^3 + 11x^2 + 68x - 60$

E. Synthetic division is a very important tool used in factoring, in finding solutions to $P(x) = 0$, and in evaluating polynomials, that is, finding the value of $P(c)$. This is a skill that is developed by practice. Remember to add $0x^k$ as a place holder when the term x^k is missing.

9. Find the quotient and remainder when $x^5 - 2x^4 - 21x^3 + 12x^2 - 3x + 5$ is divided by $x + 4$.

$x + 4 = 0$ implies $x = -4$.

$$\begin{array}{r|rrrrrr} -4 & 1 & -2 & -21 & 12 & -3 & 5 \\ & & -4 & 24 & -12 & 0 & 12 \\ \hline & 1 & -6 & 3 & 0 & -3 & \boxed{17} \end{array}$$

Quotient: $x^4 - 6x^3 + 3x^2 - 3$
Remainder: 17

10. Let $P(x) = x^5 - 3x^3 - 7x^2 - 5x + 2$. Find $P(2)$.

$$\begin{array}{r|rrrrrr} 2 & 1 & 0 & -3 & -7 & -5 & 2 \\ & & 2 & 4 & 2 & -10 & -30 \\ \hline & 1 & 2 & 1 & -5 & -15 & \boxed{-28} \end{array}$$

So $P(2) = -28$

Note: Remember to insert a 0 for the x^4 term's coefficient.

Section 5.2   Rational Roots

Key Ideas
A. Integer roots.
B. Rational roots.
C. Descartes' Rule of Signs.
D. Upper and lower bounds of roots.
E. Systematic procedure for finding all rational roots.

A. Any rational root of a polynomial with integer coefficients and leading coefficient 1 *must* be an integer that divides the constant term evenly, so that integer roots are factors of the constant term.

1. What numbers could be the integer zeros of the polynomial $x^5 + 25x^4 - 2x^3 + 121x^2 - x + 12$?	$\pm$ the factors of 12; $\pm 1, \pm 2, \pm 3, \pm 4, \pm 6, \pm 12$.
2. What numbers could be the integer zeros of the polynomial $x^7 - 8x^4 + 9x^2 - 5x + 16$?	$\pm$ the factors of 16; $\pm 1, \pm 2, \pm 4, \pm 8, \pm 16$.
3. What numbers could be the integer zeros of the polynomial $x^6 + 11x^5 + 7x^3 - 6x + 24$?	$\pm$ the factors of 24; $\pm 1, \pm 2, \pm 3, \pm 4, \pm 6, \pm 8, \pm 12, \pm 24$.

B. If $\frac{p}{q}$ is a rational root in lowest terms of the polynomial equation with *integer* coefficients $a_n x^n + a_{n-1} x^{n-1} + \cdots + a_1 x^1 + a_0 = 0$, where $a_n \neq 0$, $a_0 \neq 0$, then p is a factor of a_0 and q is a factor of a_n. That is $\frac{p}{q} = \pm \frac{\text{divisor of } a_0}{\text{divisor of } a_n}$.

4. Find all *possible* rational roots of the equation, $x^4 - x^3 - 11x^2 + 9x + 18 = 0$ and then solve it completely.

Rational roots $\pm \dfrac{\text{divisors of } 18}{\text{divisors of } 1}$
$= \pm 1, \pm 2, \pm 3, \pm 6, \pm 9, \pm 18$
There are 12 possible rational roots.
Trying the positive integers first

$$\begin{array}{r|rrrrr} 1 & 1 & -1 & -11 & 9 & 18 \\ & & 1 & 0 & -11 & -2 \\ \hline & 1 & 0 & -11 & -2 & \boxed{16} \end{array} \quad \text{no}$$

$$\begin{array}{r|rrrrr} 2 & 1 & -1 & -11 & 9 & 18 \\ & & 2 & 2 & -18 & -18 \\ \hline & 1 & 1 & -9 & -9 & \boxed{0} \end{array} \quad \text{yes}$$

At this point ± 2, ± 6, and ± 18 can be eliminated as possible roots. So we are left with only 3 and 9 as the only positive roots left.

$$\begin{array}{r|rrrr} 3 & 1 & 1 & -9 & -9 \\ & & 3 & 12 & 9 \\ \hline & 1 & 4 & 3 & \boxed{0} \end{array} \quad \text{yes}$$

Factoring the resulting polynomial yields the last two solutions.

$x^2 + 4x + 3 = (x + 1)(x + 3)$
$\quad x + 1 = 0 \qquad x + 3 = 0$
$\quad x = -1 \qquad\quad x = -3$

Solutions: $-3, -1, 2,$ and 3.

5. Find all *possible* rational roots of the equation, $10x^4 + 11x^3 - 51x^2 - 32x + 20 = 0$ and then solve it completely.

Rational roots $\pm \dfrac{\text{divisors of 20}}{\text{divisors of 10}}$

$= \pm \dfrac{1,2,4,5,10,20}{1,2,4,5,10}$

$= \pm 1, \pm 2, \pm 4, \pm 5, \pm 10, \pm 20, \pm \dfrac{1}{2}, \pm \dfrac{1}{4}, \pm \dfrac{1}{5}, \pm \dfrac{1}{10}, \pm \dfrac{2}{5}, \pm \dfrac{4}{5}, \pm \dfrac{5}{2}, \pm \dfrac{5}{4}$

There are 28 possible rational roots, many of possibilities to try. Trying the integers first (easier). Using the short cut notation from the text.

```
1 | 10   11   -51   -32    20
         10    21   -30   -62
    10   21   -30   -62 | -42      no


2 | 10   11   -51   -32    20
         20    62    22   -20
    10   31    11   -10  | 0       yes
```

Trying 2 again;
```
2 | 10   31    11   -10
         20   102
    10   51   113                  no
```

Note: *It is clear that 2 was not going to work. Also that any number larger than 2 won't work. Start trying negative numbers.*

```
-1 | 10   31    11   -10
         -10   -21    10
     10   21   -10  | 0            yes
```

Factoring this last polynomial:
$10x^2 + 21x - 10 = (2x + 5)(5x - 2)$

$2x + 5 = 0 \quad\quad 5x - 2 = 0$
$2x = -5 \quad\quad\quad 5x = 2$
$x = -\dfrac{5}{2} \quad\quad\quad x = \dfrac{2}{5}$

Solutions: $-\dfrac{5}{2}, -1, \dfrac{2}{5},$ and 2.

C. Descartes' Rule of Signs states that if $P(x)$ is a polynomial with real coefficients, then
1. The number of positive solutions of $P(x)$ is at most equal to the number variations in signs in $P(x)$ or is less than that by an even number.
2. The number of negative solutions of $P(x)$ is at most equal to the number variations in signs in $P(-x)$ or is less than that by an even number.

This is very useful when there is a odd number of variations in sign. In that case you are guaranteed at least one real solution, either negative or positive.

6. Use Descartes' Rule of Signs to help find all rational zeros of the polynomial, $P(x) = x^4 - 2x^3 - x^2 - 4x - 6$. Then solve it completely.

$P(x)$ has one variation in sign, so there is one positive real zero.
$P(-x) = x^4 + 2x^3 - x^2 + 4x - 6$ has three variation in signs, so there are either three or one negative real zeros.
These possible zeros are: $\pm 1, \pm 2, \pm 3, \pm 6$

$$\begin{array}{r|rrrrr}
-1 & 1 & -2 & -1 & -4 & -6 \\
 & & -1 & 3 & -2 & 6 \\
\hline
 & 1 & -3 & 2 & -6 & \underline{0} \quad \text{yes}
\end{array}$$

Looking at the degree three polynomial,
$Q(x) = x^3 - 3x^2 + 2x - 6$
$Q(-x) = -x^3 - 3x^2 - 2x - 6$
we can see that there is still one positive zero (we cannot increase the number of positive zeros) but *no* more negative zeros. Trying the positive numbers.

$$\begin{array}{r|rrrr}
1 & 1 & -3 & 2 & -6 \\
 & & 1 & -2 & 0 \\
\hline
 & 1 & -2 & 0 & \underline{-6}
\end{array} \quad \begin{array}{r|rrrr}
2 & 1 & -3 & 2 & -6 \\
 & & 2 & -2 & 0 \\
\hline
 & 1 & -1 & 0 & \underline{-6}
\end{array}$$

$$\begin{array}{r|rrrr}
3 & 1 & -3 & 2 & -6 \\
 & & 3 & 0 & -6 \\
\hline
 & 1 & 0 & 2 & \underline{0}
\end{array}$$

Since the result $x^2 + 2$ does not factor over the reals, we are finished. Only real solutions are -1 and 2.

D. The number a is a **lower bound** and the number b is an **upper bound** for roots of a polynomial equation if every real root c of the equation satisfies $a \leq c \leq b$. The bounds a and b are found in the following way. Let $P(x)$ be a polynomial with real coefficients.
1. If we divide $P(x)$ by $x - b$ (where $b > 0$) using synthetic division, and if the row that contains the quotient and remainder has no negative entries, then b is an upper bound for the real roots for $P(x) = 0$.
2. If we divide $P(x)$ by $x - a$ (where $a < 0$) using synthetic division, and if the row that contains the quotient and remainder has entries that are alternately nonpositive and nonnegative, then a is a lower bound for the real roots for $P(x) = 0$.

7. Show that all the real roots of the equation $2x^4 + 11x^3 + 4x^2 - 44x - 48 = 0$ lie between -6 and 3.

    $\begin{array}{r|rrrrr} -6 & 2 & 11 & 4 & -44 & -48 \\ & & -12 & 6 & -60 & 624 \\ \hline & 2 & -1 & 10 & -104 & \boxed{576} \end{array}$

    Entries alternate between nonpositive and nonnegative, therefore -6 is a lower bound.

    $\begin{array}{r|rrrrr} 3 & 2 & 11 & 4 & -44 & -48 \\ & & 6 & 51 & 165 & 363 \\ \hline & 2 & 17 & 55 & 121 & \boxed{315} \end{array}$

    Each entry and remainder is positive, therefore 3 is an upper bound.

8. Show that -1 is *not* a lower bound but -2 *is* a lower bound for the roots of the polynomial equation. Then find an integer upper bound for the roots.
    $4x^5 - 6x^3 + 7x^2 - 8x - 16 = 0$

    $\begin{array}{r|rrrrrr} -1 & 4 & 0 & -6 & 7 & -8 & -16 \\ & & -4 & 4 & 2 & -9 & 17 \\ \hline & 4 & -4 & -2 & 9 & -17 & \boxed{1} \\ & & \uparrow & & \uparrow & & \end{array}$
    Not alternating in sign.

    $\begin{array}{r|rrrrrr} -2 & 4 & 0 & -6 & 7 & -8 & -16 \\ & & -8 & 16 & -20 & 26 & -36 \\ \hline & 4 & -8 & 10 & -13 & 18 & \boxed{-52} \end{array}$
    Alternating in sign, so -2 is a lower bound.

    $\begin{array}{r|rrrrrr} 1 & 4 & 0 & -6 & 7 & -8 & -16 \\ & & 4 & 4 & -2 & 5 & 17 \\ \hline & 4 & 4 & -2 & 5 & -3 & \boxed{-19} \end{array}$

    $\begin{array}{r|rrrrrr} 2 & 4 & 0 & -6 & 7 & -8 & -16 \\ & & 8 & 16 & 20 & 54 & 92 \\ \hline & 4 & 8 & 10 & 27 & 46 & \boxed{76} \end{array}$

    So 2 is *an* upper bound, 1 is not.

E. Roots to the equation $P(x) = 0$ are the same as the zeros to the polynomial $P(x)$. To summarize these steps:
1. List all possible rational roots.
2. Use Descartes' Rule of Signs to determine the possible number of positive and negative roots.
3. Use synthetic division to test for possible roots (in order). Stop when you have reached an upper or lower bound or when all predicted positive or negative roots have been found.
4. If you find a root, repeat the process with the quotient. Remember you do not have to check roots that did not work before. If you reach a quotient that you can apply the quadratic formula to or that you can factor, then use those techniques to solve.

9. Find all rational zeros of the equation, $x^3 - 3x^2 - 10x + 24 = 0$.

1. List of possible roots:
$\{\pm 1, \pm 2, \pm 3, \pm 4, \pm 6, \pm 8, \pm 12, \pm 24\}$
2. Positive roots: two variation in sign, two or none positive roots.

Negative roots: $P(-x) = -x^3 - 3x^2 + 10x + 24$. So there are only one possible negative root.

```
1 | 1   -3   -10    24
  |      1    -2   -12
  |_____
    1   -2   -12  | 12

2 | 1   -3   -10    24
  |      2    -2   -24
  |_____
    1   -1   -12  | 0    yes
```

$x^2 - x - 12 = (x - 4)(x + 3)$
$x + 2 = 0 \quad\quad x + 4 = 0$
$x = -2 \quad\quad x = -4$

The solutions are -4, -2, $-\dfrac{3}{2}$, and 2.

10. Find all rational zeros of the equation, $2x^4 + 11x^3 + 4x^2 - 44x - 48 = 0$.

1. List of possible roots:
$$\left\{\pm\frac{1}{2}, \pm 1, \pm\frac{3}{2}, \pm 2, \pm 3, \pm 4, \pm 6, \pm 8, \pm 12, \pm 16, \pm 24, \pm 48\right\}$$

2. Positive roots: one variation in sign, one positive root.

Negative roots: $P(-x) = 2x^4 - 11x^3 + 4x^2 + 44x - 48$. So there are three or one possible negative roots.

3. Upper and lower bounds where established in problem 7 above, these are -6 and 3. So list is reduced to $\left\{\pm\frac{1}{2}, \pm 1, \pm\frac{3}{2}, \pm 2, -3, -4\right\}$

```
 1/2 | 2   11    4    -44    -48
     |      1    6      5
     | 2   12   10    -39    | no chance

  1  | 2   11    4    -44    -48
     |      2   13     17    -27
     | 2   13   17    -27    |-75

 3/2 | 2   11    4    -44    -48
     |      3    7    33/2
     | 2   14   11          No chance

  2  | 2   11    4    -44    -48
     |      4   30     68     48
     | 2   15   34     24    |0   yes

-1/2 | 2   15   34     24
     |     -1   -7   -27/2
     | 2   14   27         No chance

 -1  | 2   15   34     24
     |     -2  -13    -21
     | 2   13   21    |3

-3/2 | 2   15   34     24
     |     -3  -18    -24
     | 2   12   16    |0   yes
```

$2x^2 + 12x + 16 = 2(x^2 + 6x + 8) = 2(x + 2)(x + 4)$

$x + 2 = 0 \qquad x + 4 = 0$
$x = -2 \qquad x = -4$

The solutions are $-4, -2, -\frac{3}{2}$, and 2.

Section 5.3 Using Graphing Devices to Solve Polynomial Equations

Key Ideas
A. Locating zeros, where to look.
B. Number of possible roots.

A. There are several steps in using a graphing device to locate zeros, these are:
1. Use lower and upper bound to set the x interval for the graphing rectangle.
2. Locate possible zeros.
3. Zoom into each possible zero. Verify and approximate.

1. Use a graphing calculator to approximate the zeros to $P(x) = x^4 + 4x^3 + 2x^2 + 4x + 1$ to the nearest tenth.

    Lower bound: -4

    $$\begin{array}{r|rrrrr} -4 & 1 & 4 & 2 & 4 & 1 \\ & & -4 & 0 & -8 & 16 \\ \hline & 1 & 0 & 2 & -4 & \boxed{17} \end{array}$$

    Upper bound: 0, no sign changes implies no positive zeros.

    $[-4, 0]$ by $[-1, 1]$

    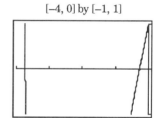

    Solutions: -3.7 and -0.3.

2. Use a graphing calculator to approximate the zeros to $P(x) = 2x^4 + 5x^3 - 3x^2 + 5x - 3$ to the nearest tenth.

    Lower bound: -6

    $$\begin{array}{r|rrrrr} -6 & 2 & 5 & -3 & 5 & -3 \\ & & -6 & 6 & -18 & 78 \\ \hline & 1 & -1 & 3 & -13 & \boxed{75} \end{array}$$

    Upper bound: 1

    $$\begin{array}{r|rrrrr} 1 & 2 & 5 & -3 & 5 & -3 \\ & & 1 & 6 & 3 & 8 \\ \hline & 1 & 6 & 3 & 8 & \boxed{5} \end{array}$$

    $[-6, 1]$ by $[-1, 1]$

    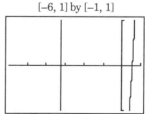

    Solutions: -3.3 and 0.6.

117

B. A polynomial of degree n can have at most n different real zeros.

3. Find all real solutions to
$3x^4 - 4x^3 + 3x^2 + x - 1 = 0$ correct to the nearest tenth.

Solutions are: $-2.5, -0.4, 0.4,$ and 0.9

[−5, 5] by [−5, 5]

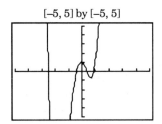

4. A company wishes to ship a tube that is 8 feet long but the air freight company will only take packages that are at most 6 feet on each side. To ship the tube the shipping department must first place it in a box that is 6 feet long and square at the ends. How wide must the square be so that the tube will fit between opposite corners of the box? Hint: The distance between opposite ends of a box is $\sqrt{\text{width}^2 + \text{thickness}^2 + \text{lenght}^2}$.

Let x = the width of the end.
This distance is $L(x) = \sqrt{x^2 + x^2 + 6^2}$.
Since x has to be positive and $x \leq 6$ (6 ft. maximum on a side) and since we want diagonal piece to be 8 ft., we use a viewing rectangle that is [0, 6] by [0,10] to graph the function $y = \sqrt{2x^2 + 36}$ and the function $y = 8$. Use the zoom and cursor to locate the point of intersection, $x = 3.74$ to two decimal places.

[0, 6] by [0,10]

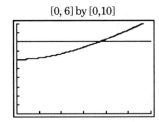

118

Section 5.4 Irrational Roots

Key Ideas
A. Intermediate value theorem.

A. The Intermediate Value Theorem is an important theorem in mathematics that helps to locate solution to polynomial equation. Basically the theorem states that if $P(a) < 0$ and $0 < P(b)$ then $P(x)$ has a solution between a and b. Repeated use of the theorem allows us to shrink the interval from a to b and narrow in to the solution to $P(x) = 0$.

1. Show that $P(x) = x^4 - 2x^3 - x^2 - 4x + 3$ has a root between 0 and 1 and a root between 2 and 3.

   Evaluate $P(x)$ at $x = 0, 1, 2, 3,$ and 4.
   $P(0) = 3$
   $P(1) = -3$
   $P(2) = -9$
   $P(3) = 9$
   Since $P(x)$ changes signs between 0 and 1 $P(x)$ has a root between 0 and 1.
   Likewise, since $P(x)$ changes signs between 2 and 3 $P(x)$ has a root between 2 and 3.

2. Show that $P(x) = x^3 + 4x^2 - 4x - 12$ has a root between 1 and 2. Find an approximate value for this zero, correct to the nearest hundred.

x	$P(x) = x^3 + 4x^2 - 4x - 12$
1	-11
2	4

   Since $P(2)$ is closer to 0, start closer to 2.

x	$P(x) = x^3 + 4x^2 - 4x - 12$
1.7	-2.327
1.8	-0.408
1.9	1.699

   Since $P(1.8)$ is closer to 0, start closer to 1.8.

x	$P(x) = x^3 + 4x^2 - 4x - 12$
1.81	-0.205
1.82	-0.001
1.83	0.204

   Solution: $x = 1.82$

3. Show that $P(x) = x^4 + 3x^3 - 5$ has exactly one positive root and one negative irrational root and find each to the nearest tenth.

Since $P(x)$ has one sign change, so there is exactly one positive root.
Since $P(-x) = x^4 - 3x^3 - 5$ has one sign change, so there is exactly one negative root.

x	$P(x) = x^4 + 3x^3 - 5$
-3	-5
-4	59
1	-1
2	35

From the evaluation above, the negative root is closer to -3 and the positive root is closer to 1. Checking for the negative root first.

x	$P(x) = x^4 + 3x^3 - 5$
-3.1	-2.021
-3.2	1.554
-3.15	-0.312
1.1	0.457
1.05	-0.312

Since $P(-3.15) < 0 < P(-3.2)$ the negative root is closer to -3.2.
Likewise, $P(1.05) < 0 < P(1.1)$ the positive root is closer to 1.1. *Remember that $P(1) = -1$.*

Section 5.5 Complex Roots and the Fundamental Theorem of Algebra

Key Ideas
A. The Fundamental Theorem of Algebra.
B. Conjugate roots.

A. This section has many important theorems and ideas. It is important to notice that these theorems *only* say that these zeroes *exists*, not *how* or *where* to find them. Sections 5.2, 5.3, and 5.4 explain *where* to look for the zeroes and *how* to find them. The **Fundamental Theorem of Algebra** states every polynomial, $P(x) = a_n x^n + a_{n-1} x^{n-1} + \cdots + a_1 x^1 + a_0$ ($n \geq 1$, $a_n \neq 0$) with complex coefficients has at least one complex zero. The **Complex Factorization Theorem** states that $P(x)$ can always be factored completely into $P(x) = a(x - c_1)(x - c_2) \cdots (x - c_{n-1})(x - c_n)$ where $a, c_1, c_2, \ldots, c_{n-1}, c_n$ are complex numbers, $a \neq 0$. If the factor $(x - c)$ appears k times in the complete factorization of $P(x)$, then c is said to have **multiplicity k**.

1. Find the complete factorization and all six zeros of the polynomial $P(x) = 5x^6 - 80x^2$.

   $P(x) = 5x^6 - 80x^2$
   $= 5x^2(x^4 - 16)$
   $= 5x^2(x^2 - 4)(x^2 + 4)$
   $= 5x^2(x - 2)(x + 2)(x - 2i)(x + 2i)$
   Set each factor equal to zero and solve.
   $5x^2 = 0$
   So $x = 0$, with multiplicity 2;
   $x - 2 = 0 \quad x + 2 = 0 \quad x - 2i = 0 \quad x + 2i = 0$
   $x = 2 \quad\quad x = -2 \quad\quad x = 2i \quad\quad x = -2i$

   The six zeroes are: 0 (multiplicity 2), -2, 2, $-2i$, $2i$

2. Find a polynomial that satisfies the given description.
   a) A polynomial $P(x)$ of degree 4 with zeros at $-2, -1, 3,$ and 5, and constant coefficient is -60.

   $P(x) = a(x - (-2))(x - (-1))(x - 3)(x - 5)$
   $= a(x + 2)(x + 1)(x - 3)(x - 5)$
   $= a(x^2 + 3x + 2)(x^2 - 8x + 15)$
   $= a(x^4 - 5x^3 - 7x^2 + 29x + 30)$
   Since $-60 = a(30)$, $a = -2$; hence
   $P(x) = -2(x^4 - 5x^3 - 7x^2 + 29x + 30)$
   $= -2x^4 + 10x^3 + 14x^2 - 58x - 60$

b) A polynomial $Q(x)$ of degree 6 with zeros at 0, 1, and -3, where 1 is zero of multiplicity 2 and -3 is zero of multiplicity 3.

$Q(x) = a(x - 0)(x - 1)^2(x - (-3))^3$
$= ax(x - 1)^2(x + 3)^3$
$= ax(x^2 - 2x + 1)(x^3 + 9x^2 + 27x + 27)$
$= ax(x^5 + 7x^4 + 10x^3 - 18x^2 - 27x + 27)$
$= a(x^6 + 7x^5 + 10x^4 - 18x^3 - 27x^2 + 27x)$

Since no information is given about Q other than its zeros, we choose $a = 1$ and we get
$Q(x) = x^6 + 7x^5 + 10x^4 - 18x^3 - 27x^2 + 27x$

c) A polynomial $R(x)$ of degree 5 with zeros at $2 - i$, $2 + i$, and 2, where 2 is zero of multiplicity 3. The leading coefficient is -1.

$R(x) = a[x - (2 - i)][x - (2 + i)][x - 2]^3$ ##
$= a[(x - 2) + i][(x - 2) - i](x - 2)^3$
$= a\left[(x^2 - 4x + 4) - i^2\right](x^3 - 6x^2 + 12x - 8)$
$= a(x^2 - 4x + 5)(x^3 - 6x^2 + 12x - 8)$
$= a(x^5 - 10x^4 + 41x^3 - 86x^2 + 92x - 40)$

Since the leading term is ax^5 and the leading term is -1, $a = -1$; hence
$R(x) = -x^5 + 10x^4 - 41x^3 + 86x^2 - 92x + 40$

Note: The difference of squares was used to multiply $[(x - 2) + i][(x - 2) - i] = (x - 2)^2 - i^2$

B. The imaginary roots of polynomial equations with real coefficients come in pairs. If the polynomial $P(x)$ of degree $n > 0$ has real coefficients, and if the complex number $a + bi$ is a root of the equation $P(x) = 0$, then so is its **complex conjugate** $a - bi$.

3. Find a polynomial $P(x)$ with real coefficient of degree 5 that has zeros at $3 - 2i$, 0, 2, and 4.

Since $3 - 2i$ is a zero, so is its conjugate, $3 + 2i$. Again we will assume $a = 1$.
So $P(x) = x[x - (3 - 2i)][x - (3 + 2i)](x - 2)(x - 4)$
$= x((x - 3) + 2i)((x - 3) - 2i)(x - 2)(x - 4)$
$= x((x^2 - 6x + 9) - 4i^2)(x^2 - 6x + 8)$
$= x(x^2 - 6x + 13)(x^2 - 6x + 8)$
$= x^5 - 12x^4 + 57x^3 - 126x^2 + 104x$

4. $-1 - 3i$ is a zero of the polynomial $P(x) = 4x^5 - 4x^4 + x^3 - 154x^2 - 158x - 40$. Find the other zeros.

Since $-1 - 3i$ is a zero of $P(x)$, $-1 + 3i$ is also a zero of $P(x)$.

$Q(x) = (x - (-1 - 3i))(x - (-1 + 3i))$
$= ((x + 1) + 3i)((x + 1) - 3i)$
$= (x + 1)^2 - (3i)^2 = x^2 + 2x + 1 - 9i^2$
$= x^2 + 2x + 10$

So $Q(x)$ is factor. Using long division;

$$\begin{array}{r} 4x^3 - 12x^2 - 15x - 4 \\ x^2 + 2x + 10 \overline{\smash{)}4x^5 - 4x^4 + x^3 - 154x^2 - 158x - 40} \\ \underline{4x^5 + 8x^4 + 40x^3} \\ -12x^4 - 39x^3 - 154x^2 \\ \underline{-12x^4 - 24x^3 - 120x^2} \\ -15x^3 - 34x^2 - 158x \\ \underline{-15x^3 - 30x^2 - 150x} \\ -4x^2 - 8x - 40 \\ \underline{-4x^2 - 8x - 40} \\ 0 \end{array}$$

Concentrating on the quotient and using the material from the previous sections.

$R(x) = 4x^3 - 12x^2 - 15x - 4$
$R(-x) = -4x^3 - 12x^2 + 15x - 4$

So $R(x)$ has exactly one positive root and two or zero negative roots. Possible rational roots are:
$\left\{ \pm \dfrac{1}{4}, \pm \dfrac{1}{2}, \pm 1, \pm 2, \pm 4 \right\}$.

Checking for the positive root first

$$\begin{array}{r|rrrr} 4 & 4 & -12 & -15 & 4 \\ & & 16 & 16 & 4 \\ \hline & 4 & 4 & 1 & \underline{|0} \end{array}$$

Factoring the resulting polynomial
$4x^2 + 4x + 1 = 0$
$(2x + 1)^2 = 0$
$2x + 1 = 0$, $2x = -1$, $x = -\dfrac{1}{2}$

So the zeros of $P(x)$ are: $-1 - 3i$, $-1 + 3i$, 4, and $-\dfrac{1}{2}$, where $-\dfrac{1}{2}$ has multiplicity 2

Section 5.6 Graphing Polynomials

Key Ideas
A. Zeros and turning points.
B. Zeroes and multiplicity.
C. Polynomial inequalities.

A. The graph of the polynomial $y = P(x)$ touches or crosses the x-axis at the point c if and only if $P(c) = 0$. Between successive zeros, the values of the polynomial are either all positive or all negative. The point at which the graph changes from increasing to decreasing or from decreasing to increasing are call **turning points** (*local extrema*). Calculus tells us that a polynomial of degree n can have at most $n - 1$ turning points.

1. Sketch the graph of the polynomial $P(x) = 2x^3 + 7x^2 - 7x - 12$.

   Start by finding the zeros. By Descartes' Rule of Sign, $P(x)$ has exactly one positive root and since $P(-x) = -2x^3 + 7x^2 + 7x - 12$, it has two or no negative roots.

   $$\begin{array}{r|rrrr} 1 & 2 & 7 & -7 & -12 \\ & & 2 & 9 & 2 \\ \hline & 2 & 9 & 2 & \boxed{-10} \end{array}$$

   $$\begin{array}{r|rrrr} 2 & 2 & 7 & -7 & -12 \\ & & 4 & 22 & 30 \\ \hline & 2 & 11 & 15 & \boxed{18} \end{array}$$

   So there is a root between 1 and 2, trying $\frac{3}{2}$;

   $$\begin{array}{r|rrrr} \frac{3}{2} & 2 & 7 & -7 & -12 \\ & & 3 & 15 & 12 \\ \hline & 2 & 10 & 8 & \boxed{0} \end{array}$$

   $2x^2 + 10x + 8 = 2(x^2 + 5x + 4) = 2(x + 1)(x + 4)$
   So the zeros are at $-4, -1, \frac{3}{2}$. Find some additional points and sketch.

   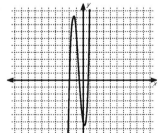

   In the graph at right, the scale is 2 to 1.

B. Multiplicity of a zero is an important tool in graphing. Suppose that c is a root of $P(x)$; If c has **even** multiplicity, then the graph of $y = P(x)$ only touches the x-axis at $x = c$. While if c has **odd** multiplicity, then the graph of $y = P(x)$ passes through the x-axis at $x = c$.

2. Sketch the graph of the polynomial $P(x) = \frac{1}{8}(x - 1)(x + 2)(x - 3)^2$.

   The polynomial $P(x)$ is degree 4. The roots –2 and 1 both have odd multiplicity, while the root 3 has even multiplicity. So the graph passes through the x-axis at $x = -2$ and at $x = 1$, but only touches the x-axis at $x = 3$. Also, $P(0) = \frac{1}{8}(-1)(2)(-3)^2 = -\frac{9}{4}$. Evaluate the polynomial at some additional points and sketch the graph.

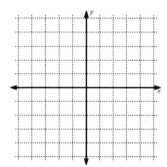

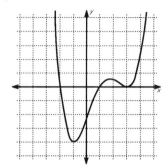

3. Sketch the graph of the polynomial $R(x) = \frac{1}{2}x^2(x + 2)^2(x + 1)(x - 2)^2$.

   The polynomial $R(x)$ is degree 7. The root –1 has odd multiplicity, while the roots –2, 0, and 2 each have even multiplicity. Since $R(0) = 0$, we to evaluate the polynomial at some additional points. Then sketch the graph.

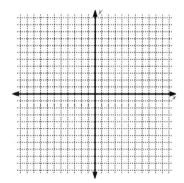

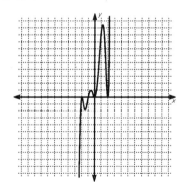

C. Follow these steps to solve a polynomial inequality.
  1. Solve the equality to generate the endpoints of the intervals.
  2. Take a test point from *within* each interval and test it in the original inequality. It makes the inequality true, then every point *within* that interval will make the inequality true (since between successive zeros, the values of the polynomial are either all positive or all negative).

It is often helpful to lay out the endpoints on a number line. This makes it easier to identify each interval correctly and to find test points from within the intervals.

4. Solve the inequality $x^2 + 4x - 21 \geq 0$.

$x^2 + 4x - 21 \geq 0$ Solve the equality.
$x^2 + 4x - 21 = 0$
$(x - 3)(x + 7) = 0$

$x - 3 = 0 \qquad x + 7 = 0$
$x = 3 \qquad x = -7$

Interval	Test Point	Test
$(-\infty, -7]$	-10	$(-10)^2 + 4(-10) - 21 \overset{?}{\geq} 0$
		$100 - 40 - 21 \overset{?}{\geq} 0$
		True
$[-7, 3]$	0	$(0)^2 + 4(0) - 21 \overset{?}{\geq} 0$
		$-21 \overset{?}{\geq} 0$
		False
$[3, \infty)$	5	$(5)^2 + 4(5) - 21 \overset{?}{\geq} 0$
		$25 + 20 - 21 \overset{?}{\geq} 0$
		True

Solution: $(-\infty, -7] \cup [3, \infty)$

5. Solve the inequality $x^3 + x^2 + 20 > 16x$.

$x^3 + x^2 + 20 > 16x$ Solve the equality.
$x^3 + x^2 + 20 = 16x$
$x^3 + x^2 - 16x + 20 = 0$
$(x + 5)(x - 2)^2 = 0$

$x - 2 = 0 \qquad x + 5 = 0$
$x = 2 \qquad x = -5$

Interval	Test Point	Test
$(-\infty, -5)$	-6	$(-6)^3 + (-6)^2 + 20 \overset{?}{>} 16(-6)$
		$-216 + 36 + 20 \overset{?}{>} -96$
		False
$(-5, 2)$	0	$(0)^3 + (0)^2 + 20 \overset{?}{>} 16(0)$
		$20 \overset{?}{>} 0$
		True
$(2, \infty)$	5	$(5)^3 + (5)^2 + 20 \overset{?}{>} 16(5)$
		$125 + 25 + 20 \overset{?}{>} 80$
		True

Solution: $(-5, 2) \cup (2, \infty)$
Note: The point 2 is not part of the solution.

Section 5.7   Using Graphing Devices to Graph Polynomials

Key Ideas
A. Local extrema of polynomials.
B. End behavior.
C. Solving inequalities.

A. A turning point is called **local maximum** when it is the highest point within some viewing rectangle and a **local minimum** when it is the lowest point *within* some viewing rectangle. A polynomial of degree n can have *at most* $n - 1$ turning points.

1. Graph the following polynomial using a graphing device. Determine the y-intercept and approximate the x-intercepts. Also approximate the coordinate of all local extrema. Approximate values to the nearest hundred.

   Pick a large viewing rectangle to make sure you find all local extrema. Then reduce the size of the viewing rectangle to find the values to the nearest hundred.

   a) $P(x) = 5x^3 - 7x^2 + x + 4$.

   y-intercept: 4 (= $P(0)$)
   x-intercept: -0.59
   local maximum: $(0.08, 4.04)$
   local minimum: $(0.86, 2.86)$

   [−5, 5] by [−5, 5]

   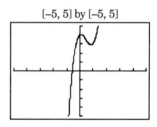

   b) $Q(x) = \frac{1}{4}x^4 - 3x^2 - 3x + 3$.

   y-intercept: 3
   x-intercepts: 0.62 and 3.79
   local maximum: $(-0.52, 3.77)$
   local minimum: $(-2.15, 0.92)$ and $(2.67, -13.69)$

   [−5, 5] by [−15, 15]

   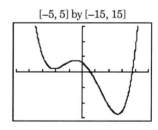

c) $R(x) = -0.1x^5 + 2x^3 - 12x - 3$.

y-intercept: −3
x-intercept: −0.25
local maximum: (−1.59, 9.06) and (3.08, −9.24)
local minimum: (−3.08, 3.24) and (1.59, −15.06)

[−5, 5] by [−20, 20]

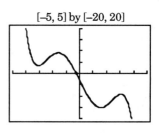

B. The **end behavior** of a function is what happens to the values of a function as $|x|$ becomes large. This means as x goes to ∞ or as x goes to $-\infty$. The end behavior of the polynomial $P(x) = a_n x^n + a_{n-1} x^{n-1} + \cdots + a_1 x^1 + a_0$ is completely determined by the leading term, $a_n x^n$. When n is even, the both ends go up when $a_n > 0$ while both ends go down for $a_n < 0$. When n is odd one end go up and the other end goes down.

2. Describe the end behavior of each polynomial.

   a) $P(x) = -2x^5 + 4x^4 - 2x^3 - 3x$.

Since you are only interested in the end behavior, use a *large* viewing rectangle.
$y \to \infty$ as $x \to -\infty$ and $y \to -\infty$ as $x \to \infty$

[−25, 25] by [−100, 100]

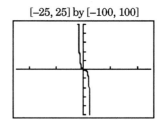

   b) $Q(x) = 0.01x^8 + 2x^7 + x + 10$.

$y \to \infty$ as $x \to -\infty$ and $y \to \infty$ as $x \to \infty$
[−25, 25] by [−500, 500]   [−50, 50] by [−500, 500]

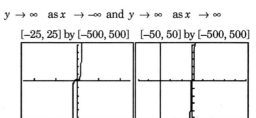

Notice the impression the viewing rectangle [−25, 25] by [−500, 500] gave was *wrong*. A larger viewing rectangle was necessary to find the true solution.

C. Inequalities of the form $f(x) > 0$ or $f(x) < 0$ can be solved by analyzing the graph of $y = f(x)$. $f(x)$ is positive when the graph is above the x-axis and $f(x)$ is negative when the graph is below the x-axis.

3. Solve the inequality by using a graphing device.

   a) $-2x^3 - 3x^2 + 11x + 6 \geq 0$

   Solution: $(-\infty, -3] \cup \left[-\dfrac{1}{2}, 2\right]$

   $[-5, 5]$ by $[-10, 10]$

   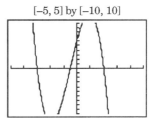

   b) $\dfrac{1}{2}x^4 - \dfrac{11}{8}x^3 - \dfrac{89}{16}x^2 - \dfrac{11}{16}x + \dfrac{15}{8} < 0$

   Solution: $\left(-2, -\dfrac{3}{4}\right) \cup \left(\dfrac{1}{2}, 5\right)$

   $[-5, 6]$ by $[-10, 10]$

   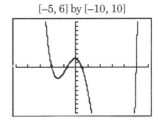

   c) $6x^2 - x - 1 \leq x^3 - x^2 + 9x - 1$

   First write the inequality as $0 \leq f(x)$.
   $6x^2 - x - 1 \leq x^3 - x^2 + 9x - 1$ Add $6x^2 - x - 1$
   $0 \leq x^3 - 7x^2 + 10$ to both sides.
   Solution: $[0, 2] \cup [5, \infty)$

   Viewing rectangle: $[-5, 5]$ by $[-10, 10]$

   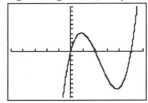

129

Section 5.8 Rational Functions

Key Ideas
A. Domain and intercepts of rational functions.
B. Vertical asymptotes and end behavior.
C. Multiplicity of factors and its effect on asymptotes.
D. Slant asymptotes.
E. Rational inequality.

A. A **rational function** is a function of the form $y = r(x) = \dfrac{P(x)}{Q(x)}$ where P and Q are polynomials. The **domain** of $r(x)$ is the set of real numbers where $Q(x) \neq 0$. The x-intercepts are the real numbers where $P(x) = 0$. The y-intercept is $r(0)$ provided 0 is in the domain of $r(x)$.

1. Find the domain, the x-intercepts, and the y-intercept of the function $r(x) = \dfrac{2x^2 + 5x + 2}{x^2 - 11x + 24}$

$r(x) = \dfrac{2x^2 + 5x + 2}{x^2 - 11x + 24}$ Factor numerator

$= \dfrac{(2x + 1)(x + 2)}{(x - 3)(x - 8)}$ and denominator.

Domain: $x \neq 3, x \neq 8$ (denominator $\neq 0$)

x–intercepts: $x = -\dfrac{1}{2}$ and $x = -2$ (numerator = 0)

y-intercept: $r(0) = \dfrac{2}{24} = \dfrac{1}{12}$

2. Find the domain, the x-intercepts, and the y-intercept of the function $t(x) = \dfrac{x^3 + 5x^2 - 6x}{x^2 - 9}$

$t(x) = \dfrac{x^3 + 5x^2 - 6x}{x^2 - 9}$ Factor numerator

$= \dfrac{x(x + 6)(x - 1)}{(x - 3)(x + 3)}$ and denominator.

Domain: $x \neq 3, x \neq -3$ (denominator $\neq 0$)
x–intercepts: $x = 0, x = 1$ and $x = -6$
                                      (numerator = 0)

y-intercept: $t(0) = \dfrac{0}{9} = 0$

B. The line $x = a$ is a **vertical asymptote** of the function $y = f(x)$ when $y \to \infty$ or $y \to -\infty$ as x approaches a from either the right side or the left side. Vertical asymptotes occur where the denominator equals zero. The end behavior of the rational function

$$r(x) = \frac{a_n x^n + a_{n-1} x^{n-1} + \cdots + a_1 x^1 + a_0}{b_m x^m + b_{m-1} x^{m-1} + \cdots + b_1 x^1 + b_0}$$ is completely determined by the leading terms

of the numerator and denominator. There are the three cases.

Case 1. $n < m$. In this case, the denominator goes to infinity faster than the numerator does. So $y \to 0$ as $x \to -\infty$ and $x \to \infty$. So $y = 0$ is a horizontal asymptote.

Case 2. $n = m$. In this case, $y \to \dfrac{a_n}{b_m}$ as $x \to -\infty$ and $x \to \infty$. So $y = \dfrac{a_n}{b_m}$ is a horizontal asymptote.

Cases 3. $n > m$. Here the end behavior is the same as the end behavior of $y = \dfrac{a_n}{b_m} x^{n-m}$.

(See *slant asymptote* below for more information about this case.)

3. Sketch the graph of the function $y = \dfrac{x^2 - 4}{x^2 - 2x - 3}$

First factor the numerator and denominator to find the domain, x-intercepts, y-intercepts, and asymptotes.

$$y = \frac{x^2 - 4}{x^2 - 2x - 3} = \frac{(x - 2)(x + 2)}{(x - 3)(x + 1)}$$

Domain: $x \neq 3$ and $x \neq -1$
Vertical asymptotes: $x = 3$ and $x = -1$
x-intercepts: $x = 2$ and $x = -2$
y-intercepts: $y = \dfrac{-4}{-3} = \dfrac{4}{3}$

Horizontal asymptote: $y = \dfrac{1}{1} = 1$

Since $n = m = 2$, both numerator and denominator have the same degree so the horizontal asymptote is the ratio of the leading coefficients

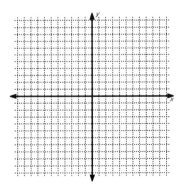

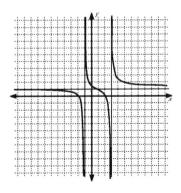

4. Sketch the graph of the function
$$y = \frac{x^3 - 3x^2 + 2x}{x^4 + 1}$$

First factor the numerator and denominator to find the domain, x-intercepts, y-intercepts, and asymptotes.
$$y = \frac{x^3 - 3x^2 + 2x}{x^4 + 1} = \frac{x(x-2)(x-1)}{x^4 + 1}$$
Domain: all real numbers, *the denominator does not factor.*
Vertical asymptotes: none
x-intercepts: $x = 0$, $x = 2$ and $x = 1$
y-intercepts: $y = \frac{0}{1} = 0$
Horizontal asymptote: $y = 0$ Since $n < m$.

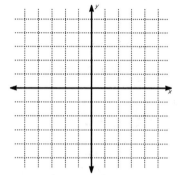

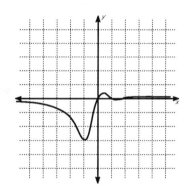

C. In the same way that multiplicity influences how the graph of a polynomial behaves around zeros, multiplicity of the zeros of the numerator and the zeros of the denominator affect the graph a rational function. If a factor occurs an even number of times, then the graph stays on the same side of the x-axis. If it occurs an odd number of times, then the graph changes sides of the x-axis at that factor.

5. Sketch the graph of the function $y = \dfrac{x^2}{x^2 - 9}$

$y = \dfrac{x^2}{x^2 - 9} = \dfrac{x^2}{(x-3)(x+3)}$
Domain: $x \neq 3$ and $x \neq -3$
Vertical asymptotes: $x = 3$ and $x = -3$. Both roots are multiplicity one, so the graph changes sign at each asymptote.
x-intercepts: $x = 0$
This is root of multiplicity two, so the graph does not change signs here.
y-intercept: $y = 0$
Horizontal asymptote: $y = 1$

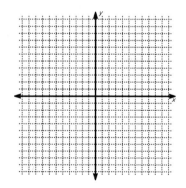

132

6. Sketch the graph of the function $y = \dfrac{x^2 - 9}{x^2}$

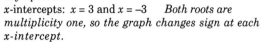

Domain: $x \neq 0$
Vertical asymptotes: $x = 0$ *This is a multiplicity two root, so the graph does not change signs at this asymptote.*
x-intercepts: $x = 3$ and $x = -3$ *Both roots are multiplicity one, so the graph changes sign at each x-intercept.*
y-intercept: None, 0 is *not* in the domain.
Horizontal asymptote: $y = 1$

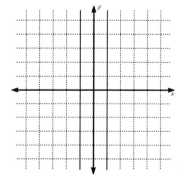
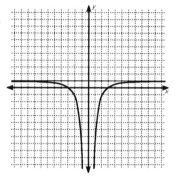

D. When $n > m$ in $r(x) = \dfrac{P(x)}{Q(x)} = \dfrac{a_n x^n + a_{n-1} x^{n-1} + \cdots + a_1 x^1 + a_0}{b_m x^m + b_{m-1} x^{m-1} + \cdots + b_1 x^1 + b_0}$, we use long division to write $r(x) = D(x) + \dfrac{R(x)}{Q(x)}$ where the degree of $R(x) <$ the degree of $Q(x)$. The polynomial $D(x)$ is a **slant** or **oblique asymptote**. This means that as $|x|$ get large, the graph of $y = r(x)$ behaves like the graph of $y = D(x)$.

7. Sketch the graph of the function $y = \dfrac{x^3 - 4x}{x^2 - 1}$

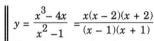

$y = \dfrac{x^3 - 4x}{x^2 - 1} = \dfrac{x(x-2)(x+2)}{(x-1)(x+1)}$

Domain: $x \neq -1$ and $x \neq 1$
Vertical asymptotes: $x = -1$ and $x = 1$ *Both multiplicity one.*
x intercepts: $x = 0$, $x = 2$, and $x = -2$ *All three are multiplicity one.*
y-intercept: $y = 0$
Slant asymptote: $y = x$ since
$y = \dfrac{x^3 - 4x}{x^2 - 1} = x - \dfrac{3x}{x^2 - 1}$

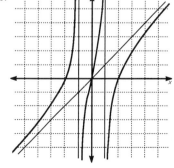

133

E. As in the case with polynomials and polynomial inequalities, rational functions can only change its sign at a finite number of places. These places are the zeros of the equality and zeroes of each denominator. The steps are basically the same as in polynomials. First find these places and then test a point from *within* each interval in the original inequality. Use a number line to help in defining the intervals.

8. Solve the inequality $\dfrac{x-2}{x+3} \geq 0$

$\dfrac{x-2}{x+3} = 0 \implies x - 2 = 0$ Solve the equality.
$x = 2$
Denominator is 0 when $x + 3 = 0$, or when $x = -3$

Interval	Test Point	Test	
$(-\infty, -3)$	-10	$\dfrac{(-10)-2}{(-10)+3} \overset{?}{\geq} 0$	
		$\dfrac{-12}{-7} = \dfrac{12}{7} \overset{?}{\geq} 0$	True
$(-3, 2]$	0	$\dfrac{(0)-2}{(0)+3} \overset{?}{\geq} 0$	
		$\dfrac{-2}{3} \overset{?}{\geq} 0$	False
$[2, \infty)$	5	$\dfrac{(5)-2}{(5)+3} \overset{?}{\geq} 0$	
		$\dfrac{3}{8} \overset{?}{\geq} 0$	True

Solution: $(-\infty, -3) \cup [2, \infty)$ $x = -3$ *is not included in any interval since it is not in the domain.*

9. Solve the inequality $\dfrac{x-3}{2x+1} < \dfrac{5}{3}$

$\dfrac{x-3}{2x+1} = \dfrac{5}{3}$ Solve the equality.
$3x - 9 = 10x + 5$ Cross multiply.
$-7x = 14$ So $x = -2$ is a zero.
Denominator: $2x + 1 = 0 \implies x = -\dfrac{1}{2}$

Interval	Test Point	Test	
$(-\infty, -2]$	-10	$\dfrac{(-10)-3}{2(-10)+1} \overset{?}{<} \dfrac{5}{3}$	
		$\dfrac{-13}{-19} \overset{?}{<} \dfrac{5}{3}$	True
$\left[-2, -\dfrac{1}{2}\right)$	-1	$\dfrac{(-1)-3}{2(-1)+1} \overset{?}{<} \dfrac{5}{3}$	
		$\dfrac{-4}{-1} \overset{?}{<} \dfrac{5}{3}$	False
$\left(-\dfrac{1}{2}, \infty\right)$	0	$\dfrac{(0)-3}{2(0)+1} \overset{?}{<} \dfrac{5}{3}$	
		$\dfrac{-3}{1} \overset{?}{<} \dfrac{5}{3}$	true

Solution: $(-\infty, -2] \cup \left(\dfrac{1}{2}, \infty\right)$

10. Solve the inequality $\dfrac{x}{x+3} < \dfrac{x+1}{x-3}$

$\dfrac{x}{x+3} < \dfrac{x+1}{x-3}$ Solve the equality.

$\dfrac{x}{x+3} = \dfrac{x+1}{x-3}$

$\dfrac{x(x-3)}{(x+3)(x-3)} = \dfrac{(x+1)(x+3)}{(x-3)(x+3)}$

$\dfrac{x^2-3x}{(x+3)(x-3)} = \dfrac{x^2+4x+3}{(x+3)(x-3)}$

$\dfrac{x^2-3x}{(x+3)(x-3)} - \dfrac{x^2+4x+3}{(x+3)(x-3)} = 0$

$\dfrac{-7x-3}{(x+3)(x-3)} = 0$

$-7x-3 = 0$ implies $x = -\dfrac{3}{7}$

Denominator: $x+3=0$ or $x-3=0$
            $x=-3$ $x=3$

Interval	Test Point	Test	
$(-\infty, -3)$	-5	$\dfrac{(-5)}{(-5)+3} \overset{?}{<} \dfrac{(-5)+1}{(-5)-3}$	
		$\dfrac{-5}{-2} \overset{?}{<} \dfrac{-4}{-8}$	
		$\dfrac{5}{2} \overset{?}{<} \dfrac{1}{2}$	False
$\left(-3, -\dfrac{3}{7}\right)$	-2	$\dfrac{(-2)}{(-2)+3} \overset{?}{<} \dfrac{(-2)+1}{(-2)-3}$	
		$\dfrac{-2}{1} \overset{?}{<} \dfrac{-1}{-5}$	
		$-2 \overset{?}{<} \dfrac{1}{5}$	True
$\left(-\dfrac{3}{7}, 3\right)$	0	$\dfrac{(0)}{(0)+3} \overset{?}{<} \dfrac{(0)+1}{(0)-3}$	
		$0 \overset{?}{<} \dfrac{1}{-3}$	False
$(3, \infty)$	5	$\dfrac{(5)}{(5)+3} \overset{?}{<} \dfrac{(5)+1}{(5)-3}$	
		$\dfrac{5}{8} \overset{?}{<} \dfrac{6}{2} = 3$	True

Solution: $\left(-3, -\dfrac{3}{7}\right) \cup (3, \infty)$

Section 5.9 Using Graphing Devices to Graph Rational Functions

Key Ideas
A. Vertical asymptotes and end behavior on graphing devices.

A. Some graphing calculators and computer graphing programs do not graph function with vertical asymptotes properly. Sometimes vertical asymptotes will appear as vertical lines. Other problems are caused by viewing rectangles that are too wide. In this case, vertical asymptotes might disappear altogether. However, wide viewing rectangles are useful in showing horizontal asymptotes as well as other end behavior. The key is to use a variety of sizes of viewing rectangles to determine both vertical asymptotes as well as the end behavior.

1. Graph each rational function in the given viewing rectangles. Determine all vertical and horizontal asymptotes from the graphs.

   a) $y = \dfrac{5x + 7}{x - 3}$ [−5, 5] by [−20, 20]
   [−50, 50] by [−1, 10]

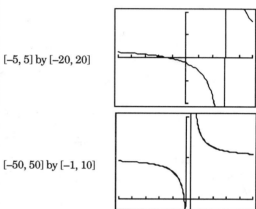

Vertical asymptotes: $x = 3$
Horizontal asymptotes: $y = 5$

   b) $y = \dfrac{x^2 - 4}{x^3 + 3x}$ [−5, 5] by [−5, 5]
   [−20, 20] by [−5, 5]

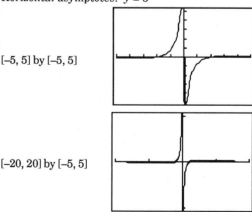

Vertical asymptotes: $x = 0$
Horizontal asymptotes: $y = 0$

c) $y = \dfrac{8 + 3x + 2x^2}{9 - x^2}$ [−5, 5] by [−5, 5]

[−50, 50] by [−5, 5]

[−5, 5] by [−5, 5]

[−50, 50] by [−5, 5]

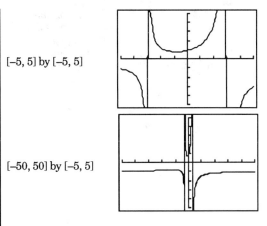

Vertical asymptotes: $x = -3$ and $x = 3$
Horizontal asymptotes: $y = -2$

2. Graph the rational function in appropriate viewing rectangles. Determine all vertical and horizontal asymptotes, its x and y-intercepts and all local entrema correct to two decimal places.
$$y = \dfrac{3x^2 + x - 4}{x^2 - 5x + 6}$$

[−10, 10] by [−10, 10]

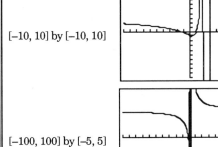

[−100, 100] by [−5, 5]

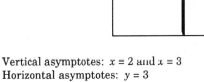

Vertical asymptotes: $x = 2$ and $x = 3$
Horizontal asymptotes: $y = 3$
x-intercept: $x = -1.33$ and $x = 1.00$
y-intercept: $y = -0.67$
Extreme point is at $(0.37, -0.75)$

[0, 1] by [−2, 0]

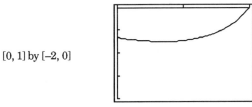

3. Graph each rational function in appropriate viewing rectangles. Determine all vertical asymptotes. Then graph f and g in a sufficiently large viewing rectangle to show that they have the same end behavior.

   a) $f(x) = \dfrac{x^3 + 7x^2 - 5}{x^2 - 1}$

   $g(x) = x + 7$

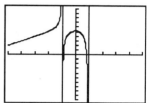

[−5, 5] by [−10, 10]

Vertical asymptotes: $x = -1$ and $x = 1$

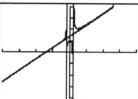

[−20, 20] by [−20, 20]

$f(x) = x + 7 + \dfrac{x + 2}{x^2 - 1}$

   b) $f(x) = \dfrac{x^3 - x + 12}{x + 4}$

   $g(x) = x^2 - 4x + 15$

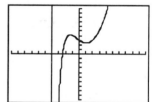

[−10, 10] by [−10, 10]

Vertical asymptote: $x = -4$

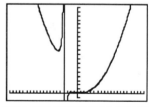

[−20, 20] by [−10, 200]

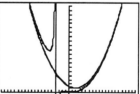

[−20, 20] by [−10, 200]

$f(x) = x^2 - 4x + 15 - \dfrac{48}{x + 4}$

Chapter 6

Exponential and Logarithmic Functions

Section 6.1 Exponential Functions
Section 6.2 Application: Exponential Growth and Decay
Section 6.3 Logarithmic Functions
Section 6.4 Laws of Logarithms
Section 6.5 Applications of Logarithms

Section 6.1   Exponential Functions

Key Ideas
A. Exponential functions and their graphs.
B. The exponential function with base e.

A. For $a > 0$, the **exponential function with base a** is defined as $f(x) = a^x$ for every real number x. The value of the base determines the general shape of the graph of $f(x) = a^x$. For $0 < a < 1$, the graph of $f(x) = a^x$ decreases rapidly. When $a = 1$, $f(x) = 1^x = 1$ is a constant function. And for $a > 1$, the function $f(x) = a^x$ increases rapidly. Remember: $a^{-1} = \dfrac{1}{a}$, $a^0 = 1$, and a positive number to any power is always positive. As a result, the domain of $f(x) = a^x$ is the real numbers and the range is the positive real numbers.

1. Sketch the graphs of $y = 2^x$ and $y = \left(\dfrac{1}{2}\right)^x$ on the same axis.

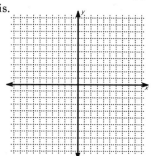

Since $\dfrac{1}{2} = 2^{-1}$, $y = \left(\dfrac{1}{2}\right)^x = 2^{-x}$.

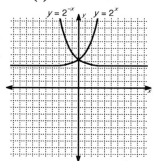

2. Sketch the graph of $y = 3^x$ and use it to sketch the graphs of $y = -2 + 3^x$ and $y = 4 - 3^x$ (all on the same axis).

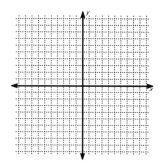

First graph $y = 3^x$.
To graph $y = -2 + 3^x$ start with $y = 3^x$ and shift the graph vertically 2 units down.
To graph $y = 4 - 3^x$ start with $y = 3^x$, first flip the graph of $y = 3^x$ across the x-axis, and *then* shift the result 4 units up.

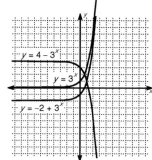

B. As n becomes large, the values of $\left(1+\dfrac{1}{n}\right)^n$ approaches an irrational number called e, $e \approx 2.71828$. The function $f(x) = e^x$ is called the **natural exponential function**. Its domain is the real numbers and its range is the positive real numbers.

3. Solve $x^5 e^x + 3x^4 e^x = 4x^3 e^x$.

$x^5 e^x + 3x^4 e^x = 4x^3 e^x$
$x^5 e^x + 3x^4 e^x - 4x^3 e^x = 0$
$\left(x^5 + 3x^4 - 4x^3\right) e^x = 0$ *Factor out the e^x.*
$x^3\left(x^2 + 3x - 4\right) e^x = 0$
$x^3 = 0 \qquad x^2 + 3x - 4 = 0 \qquad e^x \neq 0$ *Since $e^x > 0$.*
$x = 0 \qquad (x+4)(x-1) = 0$
$\qquad\qquad x+4=0 \qquad x-1=0$
$\qquad\qquad x=-4 \qquad x=1$
Solution: $\{0, -4, 1\}$

4. Use a graphing calculator to solve $e^{2x} = 5e^x$ to two decimal places.

First set equal to zero:
$e^{2x} = 5e^x \qquad e^{2x} - 5e^x = 0$
Now graph $f(x) = e^{2x} - 5e^x$ and find where it intersects the x-axis.
Solution: $\{1.61\}$

[−2,2] by [−2, 2]

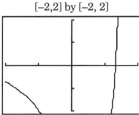

Note: A larger viewing rectangle shows that $f(x)$ approaches the x-axis, but it does not touch the axis.

5. Use a graphing calculator to solve $xe^x - 3x^3 = 0$ to two decimal places.

Graphing $f(x) = xe^x - 3x^3$ and find where it intersects the x-axis.
Solution: $\{-0.46, 0, 0.91, 3.73\}$

[−1,4] by [−2, 2]

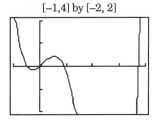

Section 6.2   Application: Exponential Growth and Decay

Key Ideas
A. Exponential growth.
B. Radioactive decay.
C. Compound interest.

A. If d is the time required for a population to double, the **doubling period**, then the size of the population at time t is given by $n(t) = n_0 2^{\frac{t}{d}}$, where n_0 is the initial size of the population.

1. Under ideal conditions a certain bacteria population is known to double in size every $4\frac{1}{2}$ hours. Suppose that there are initially 2000 bacteria.
   a) What will be the size of the population in 18 hours?

   $n(t) = n_0 2^{\frac{t}{d}} = (2000)\, 2^{\frac{18}{(9/2)}}$
   $= (2000)\, 2^4$
   $= 32000$

   b) What is the size after t hours?

   $n(t) = n_0 2^{\frac{t}{d}} = (2000)\, 2^{\frac{t}{(9/2)}}$
   $= (2000)\, 2^{\frac{2t}{9}}$

2. A biologist finds that a population of yeast doubles in size every 45 minutes. The estimated yeast count after 3 hours is 56,000,000.
   a) What was the initial size of the yeast culture?

   Here we want to find $n(0)$. First convert to a common unit of time, here we chose minutes.
   3 hours = 180 minutes

   $n(t) = n_0 2^{\frac{t}{d}}$
   Here $d = 45$ minutes and
   $n(180) = 56000000$

   $n(180) = n_0 2^{\frac{180}{45}}$
   $56000000 = n_0\, 2^4 = n_0\, 16$
   So, $n_0 = \frac{56000000}{16} = 3{,}500{,}000$

b) What is the estimated size after $7\frac{1}{2}$ hours?

Convert hours to minutes. $7\frac{1}{2}$ hours = 450 minutes
Using $n_0 = 3{,}500{,}000$, and $d = 45$ minutes

$$n(t) = n_0 2^{\frac{t}{d}}$$

$$n(450) = (3{,}500{,}000) 2^{\frac{450}{45}} = (3{,}500{,}000) 2^{10}$$
$$= 3{,}584{,}000{,}000$$

c) What is the estimated size after 2 days?

Convert days to minutes.
2 days = 48 hours = 2880 minutes
Using $n_0 = 3{,}500{,}000$, and $d = 45$ minutes

$$n(t) = n_0 2^{\frac{t}{d}}$$

$$n(2880) = (3{,}500{,}000) 2^{\frac{2880}{45}} = (3{,}500{,}000) 2^{64}$$
$$= 6.61 \times 10^{28}$$

B. The **half-life** of radioactive material is the period of time during which half the mass of the radioactive material disintegrates. The mass of the material that remains after time t is given by $m(t) = m_0 2^{-\frac{t}{h}}$, where m_0 is the initial mass and h is the half-life.

3. Carbon-14 has a half-life of 5730 years. If a sample has a mass of 150 mg, what would the mass of the sample be in 100 years?

$$m(t) = m_0 \left(2^{-\frac{t}{h}} \right)$$

$$m(t) = (150) \left(2^{-\frac{t}{5730}} \right)$$

Setting $t = 100$, we get;

$$m(100) = (150) \left(2^{-\frac{100}{5730}} \right) = 148.196 \text{ mg}$$

4. Einsteinium-254 has a half life of 276 days. One year later a sample of Einsteinium-254 has a mass of 86 mg. Find the initial mass.

First convert years to days and use $t = 365$ and $m(365) = 86$ to solve for m_0.

$$m(t) = m_0 \left(2^{-\frac{t}{276}} \right)$$

$$m(365) = m_0 \left(2^{-\frac{365}{276}} \right)$$

$$86 = m_0 \left(2^{-1.32246} \right) = m_0 (0.399985)$$

$$m_0 = \frac{86}{0.399985} = 215.080 \text{ mg}$$

C. If an amount of money, P, called the **principal**, is invested at a rate of r, compounded n **compounding periods** per year, then the amount of money after t years is given by $A = P\left(1 + \frac{r}{n}\right)^{nt}$. The rate $\frac{r}{n}$ is the *rate per compounding period* and nt is the number of compounding periods. As the number of compounding periods per year become large, the value of the amount approaches $A = Pe^{rt}$ called **continuous compounding of interest**.

5. Compare the amount after 5 years when $1000 is invested in each type of account.

    a) 10% compounded monthly.

    $P = 1000, r = 0.1, n = 12, t = 5$
    $A = P\left(1 + \frac{r}{n}\right)^{nt}$
    $A = 1000\left(1 + \frac{0.1}{12}\right)^{(12)(5)} = 1000(1.008333333)^{60}$
    $= 1000(1.645308934) = 1645.31$
    $A = \$1645.31$

    b) 9.9% compounded weekly, assume 52 weeks per year.

    $P = 1000, r = 0.099, n = 52, t = 5$
    $A = P\left(1 + \frac{r}{n}\right)^{nt}$
    $A = 1000\left(1 + \frac{0.099}{52}\right)^{(52)(5)} = 1000(1.001903846)^{260}$
    $= 1000(1.639726393) = 1639.26$
    $A = \$1639.26$

    c) 9.9% compounded continuously.

    $P = 1000, r = 0.099, t = 5$
    $A = Pe^{rt}$
    $A = 1000e^{(0.099)(5)} = 1000e^{0.495}$
    $= 1000(1.640498239) = 1640.50$
    $A = \$1640.50$

6. How much money needs to be invested in an account paying 7.8% compounded monthly, so that a person will have $5000 in 5 years?

    Solve for P.
    $A = P\left(1 + \frac{r}{n}\right)^{nt}$ So $P = \dfrac{A}{\left(1 + \frac{r}{n}\right)^{nt}}$

    Using $A = 5000, r = 0.078, n = 12, t = 5$
    $P = \dfrac{5000}{\left(1 + \frac{0.078}{12}\right)^{(12)(5)}} = \dfrac{5000}{(1.0065)^{60}}$
    $P = \$3389.56$

Section 6.3   Logarithmic Functions

Key Ideas
A. What is a logarithm?
B. Graphs of logarithmic functions.
C. Natural logarithms.

A. For $a \neq 1$, the inverse of the exponential function $f(x) = a^x$ is a function called a **logarithm** and written as $g(x) = \log_a x$. $\log_a x$ is read as '*log base a of x*'. The statement $\log_a x = y$ is equivalent to the statement $a^y = x$. $\log_a 1 = 0$ and $\log_a a = 1$. $\log_a x$ is the exponent to which the base a must be raised to give x.

1. Evaluate.
   a) $\log_5 25$

   $\log_5 25 = 2$
   since $5^2 = 25$

   b) $\log_2 \frac{1}{8}$

   $\log_2 \frac{1}{8} = -3$
   since $2^{-3} = \frac{1}{8}$

   c) $\log_{(\frac{1}{3})} 27$

   $\log_{(\frac{1}{3})} 27 = -3$
   since $\left(\frac{1}{3}\right)^{-3} = 3^3 = 27$

2. Solve for x. $\log_5 (62 - x) = 3$

   $\log_5 (62 - x) = 3$
   $62 - x = 5^3$
   $62 - x = 125$
   $-x = 63$
   $x = -63$

   Check:
   $\log_5 (62 - (-63)) = \log_5 125 = \log_5 5^3 = 3$ ✓

3. Solve for x. $2^{3x+1} = 32$

$2^{3x+1} = 32$
$\log_2(2^{3x+1}) = \log_2 32$
$(3x+1)\log_2 2 = \log_2 2^5$
$3x + 1 = 5$
$3x = 4$
$x = \dfrac{4}{3}$

B. The domain for $f(x) = \log_a x$ is the positive real numbers and the range is the real numbers. As with exponential functions, the graph of $y = \log_a x$ is directly depended on the value of a. The graphs of $f(x) = \log_a x$ and $f^{-1}(x) = a^x$ for various values of a are shown below.

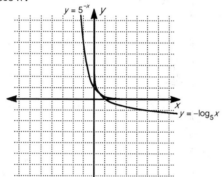
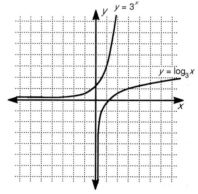

4. Determine the domain and range for $f(x) = \log_3(x - 2)$. Sketch the graph of $y = f(x)$.

Domain: $x - 2 > 0$
$ x > 2$
Range: $(-\infty, \infty)$

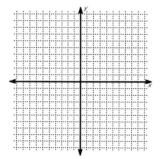
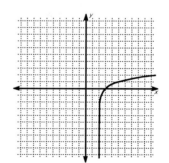

5. Determine the domain and range for $f(x) = \log_5(x^2 + 1)$. Sketch the graph of $y = f(x)$.

Domain: Since $x^2 + 1 \geq 1$ for all x, the domain is $(-\infty, \infty)$.
Also since, $x^2 + 1 \geq 1$, $\log_5(x^2 + 1) \geq \log_5 1 = 0$, so the Range is $[0, \infty)$, the nonnegative reals.

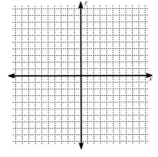
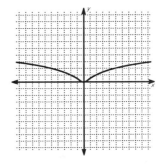

C. The logarithm with base e is called the **natural logarithm** and is given a special name and notation, $\log_e x = \ln x$. The logarithm with base 10 is called **common logarithm** and given an abbreviated notation, $\log_{10} x = \log x$.

$$\ln x = y \quad \Leftrightarrow \quad e^y = x \qquad \text{and} \qquad \log x = y \quad \Leftrightarrow \quad 10^y = x$$

6. Solve the equation $4e^{2x-1} = 7$.

$e^{2x-1} = \frac{7}{4}$

$\ln e^{2x-1} = 2x - 1 = \ln\left(\frac{7}{4}\right)$

$2x = 1 + \ln\left(\frac{7}{4}\right)$

$x = \frac{1}{2} + \frac{1}{2}\ln\left(\frac{7}{4}\right) \approx 1.06$

7. Solve the equation $e^{2x} - 4e^x + 3 = 0$.

Factor. $e^{2x} - 4e^x + 3 = 0$
$(e^x - 1)(e^x - 3) = 0$
$e^x - 1 = 0 \qquad\qquad e^x - 3 = 0$
$e^x = 1 \qquad\qquad\quad\ e^x = 3$
$\ln(e^x) = \ln 1 = 0 \qquad \ln(e^x) = \ln 3$
So $x = 0$ or $x = \ln 3$

8. Find the domain of $f(x) = \ln(x+2) + \ln(3-2x)$. Draw the graph of $y = f(x)$ on a calculator and use it to find the asymptotes and local maximums and local minimums.

Since the input to a logarithm function must be positive;

$x + 2 > 0$ and $3 - 2x > 0$
$x > -2$ $3 > 2x$
$-2 < x$ $\frac{3}{2} > x$
$-2 < x$ $x < \frac{3}{2}$

So, domain : $-2 < x < \frac{3}{2}$

Use the domain for the view rectangle:

[−2, 1.5] by [−8, 2]

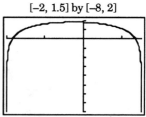

Vertical asymptotes are at $x = -2$ and at $x = \frac{3}{2}$.

Local maximum at (−0.25, 1.812); no local minimum.

Section 6.4 Laws of Logarithms

Key Ideas
A. Basic laws of logarithms.
B. Common errors to avoid!
C. Common logarithms.
D. Change of base.

A. Suppose that $x > 0$, $y > 0$, and r is any real number, then the Laws of Logarithms are :
1. $\log_a(xy) = \log_a x + \log_a y$ *The log of a product is the sum of the logs.*
2. $\log_a\left(\dfrac{x}{y}\right) = \log_a x - \log_a y$ *The log of a quotient is the difference of the logs.*
3. $\log_a(x^r) = r\log_a x$

1. Use the Laws of Logarithms to rewrite the following expression in a form that contains no logarithms of products, quotients, or powers.

a) $\log_2 9x$

$\log_2 9x = \log_2 9 + \log_2 x$
The log of a product is the sum of the logs.

b) $\log_5\left(\dfrac{2x}{y}\right)$

$\log_5\left(\dfrac{2x}{y}\right) = \log_5(2x) - \log_5 y$
$= \log_5 2 + \log_5 x - \log_5 y$

c) $\log_3 x^4 y^3$

$\log_3 x^4 y^3 = \log_3 x^4 + \log_3 y^3$
$= 4\log_3 x + 3\log_3 y$

d) $\log_4\left(\dfrac{16w^5}{\sqrt[3]{x^2 y}}\right)$

$\log_4\left(\dfrac{16w^5}{\sqrt[3]{x^2 y}}\right) = \log_4(16w^5) - \log_4\left(\sqrt[3]{x^2 y}\right)$
$= \log_4 16 + \log_4 w^5 - \dfrac{1}{3}\log_4(x^2 y)$
$= \log_4 16 + \log_4 w^5 - \dfrac{2}{3}\log_4 x - \dfrac{1}{3}\log_4 y$
$= 2 + 5\log_4 w - \dfrac{2}{3}\log_4 x - \dfrac{1}{3}\log_4 y$

2. Use the Laws of Logarithms to rewrite each of the following as a single logarithm.

a) $4\log_3 x - \frac{3}{7}\log_3 y$

$$4\log_3 x - \frac{3}{7}\log_3 y = \log_3 x^4 - \log_3 \sqrt[7]{y^3}$$

$$= \log_3\left(\frac{x^4}{\sqrt[7]{y^3}}\right)$$

b) $2\log_{10} x + 3\log_{10} y - \frac{1}{2}\log_{10} 8$

$$2\log_{10} x + 3\log_{10} y - \frac{1}{2}\log_{10} 8$$

$$= \log_{10} x^2 + \log_{10} y^3 - \log_{10} \sqrt{8}$$

$$= \log_{10}\left(\frac{x^2 y^3}{\sqrt{8}}\right)$$

c) $4\log_5 x + 2 - \log_5 y - 2\log_5 w$

$$4\log_5 x + 2 - \log_5 y - 2\log_5 w$$

$$= \log_5 x^4 + \log_5 25 - \log_5 y - \log_5 w^2$$

$$= \log_5\left(\frac{25x^4}{y w^2}\right)$$

B. There are several common mistakes that you need to avoid, these are:

1. $\log_a(x + y) \neq \log_a x + \log_a y \qquad \log_a(x + y)$ cannot be simplified!

2. $\dfrac{\log_a x}{\log_a y} \neq \log_a x - \log_a y$

3. $(\log_a x)^r \neq r\log_a x \qquad\qquad$ Here the exponent is applied to the entire function $\log_a x$, not just the input x.

3. Solve the equation $\log_8(x + 2) + \log_8(x + 5) = \frac{2}{3}$

$\log_8(x + 2) + \log_8(x + 5) = \frac{2}{3}$

$\log_8(x + 2)(x + 5) = \frac{2}{3}$

$\log_8\left(x^2 + 7x + 10\right) = \frac{2}{3}$

$8^{\log_8(x^2 + 7x + 10)} = 8^{\left(\frac{2}{3}\right)} = 4$

$x^2 + 7x + 10 = 4$
$x^2 + 7x + 6 = 0$
$(x + 6)(x + 1) = 0$
$x + 6 = 0 \quad x = -6$
$x + 1 = 0 \quad x = -1$
Check:
$x = -6$:

$\log_8((-6) + 2) + \log_8((-6) + 5) \stackrel{?}{=} \frac{2}{3}$

$\log_8(-4)$ and $\log_8(-1)$ are not defined, so $x = -6$ is NOT a solution.
$x = -1$:

$\log_8((-1) + 2) + \log_8((-1) + 5) \stackrel{?}{=} \frac{2}{3}$

$\log_8 1 + \log_8 4 = \log_8 4 = \frac{2}{3}$ ✓

Solution: $\{-1\}$

4. Solve the equation $2\log_5(x + 8) = \log_5(5x + 36)$

$2\log_5(x + 8) = \log_5(5x + 36)$

$\log_5(x + 8)^2 = \log_5(5x + 36)$

$(x + 8)^2 = 5x + 36$
$x^2 + 16x + 64 = 5x + 36$
$x^2 + 11x + 28 = 0$
$(x + 4)(x + 7) = 0$
$x + 4 = 0, \quad x = -4$
$x + 7 = 0, \quad x = -7$

$2\log_5(((-4) + 8) \stackrel{?}{=} \log_5(5(-4) + 36)$

$2\log_5 4 \stackrel{?}{=} \log_5 16$, yes, since $\log_5 4^2 = \log_5 16$,

$2\log_5 1 \stackrel{?}{=} \log_5 1$, yes, since $\log_5 1 = 0$

C. Logarithms with base 10 are called **common logarithms** and are denoted by omitting the base: $\log_{10} x = \log x$. Remember, logarithms with base e are called **natural logarithms** and $\log_e x = \ln x$.

5. Find the solution to the equation $3^{x+1} = 4$ correct to six decimal places.

$3^{x+1} = 4$
$\log 3^{x+1} = \log 4$
$(x+1)\log 3 = \log 4$
$x + 1 = \dfrac{\log 4}{\log 3}$
$x = \dfrac{\log 4}{\log 3} - 1 \approx 0.261860$

6. Find the solution to the equation $4^{2x-3} = 7^{x+1}$ correct to six decimal places.

$4^{2x-3} = 7^{x+1}$
Take log of both sides:
$(2x - 3)\log 4 = (x + 1)\log 7$
Distribute.
$2x \log 4 - 3 \log 4 = x \log 7 + \log 7$
Collect x on one side;
$2x \log 4 - x \log 7 = \log 7 + 3 \log 4$
$(2 \log 4 - \log 7)x = \log 7 + 3 \log 4$
$x = \dfrac{\log 7 + 3 \log 4}{2 \log 4 - \log 7} \approx 7.384724$

D. Since most calculators only have the common log and natural log functions, it is important to be able to change the base of a logarithm. The **change of base formula** $\log_b x = \dfrac{\log_a x}{\log_a b}$. It allows us to express a logarithm to any base (provided the base is greater than 0 and not equal to 1). The change of base formula can also be expressed as $\log_a b \cdot \log_b x = \log_a x$. Of particular importance are: $\log_b x = \dfrac{\log x}{\log b}$ or $\log_b x = \dfrac{\ln x}{\ln b}$.

7. Evaluate $\log_6 7$ to six decimal places.

Using common logs:
$\log_6 7 = \dfrac{\log 7}{\log 6} \approx \dfrac{0.84509804}{0.778151250} \approx 1.086033$
Using natural logs:
$\log_6 7 = \dfrac{\ln 7}{\ln 6} \approx \dfrac{1.945910149}{1.791759469} \approx 1.086033$

8. Simplify $(\log_3 8)(\log_8 7)$.

$(\log_3 8)(\log_8 7) = \log_3 7$

9. Solve for x. $\log_3 x + \log_9 x = 6$

We need to convert to a common base. Since $9 = 3^2$, we convert $\log_3 x$ to base 9. Since $(\log_9 3)(\log_3 x) = \log_9 x$ and $\log_9 3 = \frac{1}{2}$;

$\left(\dfrac{\log_9 3}{\frac{1}{2}}\right)(\log_3 x) + \log_9 x = 6$

$2\log_9 x + \log_9 x = 6$

$3\log_9 x = 6 \qquad \log_9 x = 2$

So $x = 9^2 = 81$

Checking:

$\log_3(81) + \log_9(81) \stackrel{?}{=} 6$

$\log_3(81) + \log_9(81) = \log_3(3^4) + \log_9(9^2)$

$= 4 + 2 = 6$, checks.

Solution: {81}

Section 6.5 Applications of Logarithms

Key Ideas
A. Using logarithms to solve exponential growth and radioactive decay problems.
B. Newton's Law of Cooling and compound interest.
C. Logarithmic scales: pH, Ricther, decibel scales.

A. Logarithms are used to extract the power in exponential functions. This is an especially useful tool when the exponent contains the unknown variable.

1. At a certain time, a bacterial culture has 115,200 bacteria. Four hours later the culture has 1,440,000 bacteria. Find the doubling time.

   $n(t) = n_0 2^{\frac{t}{d}}$ solve for d.

   $\ln n(t) = \ln\left(n_0 2^{\frac{t}{d}}\right)$

   $\ln n(t) = \ln n_0 + \ln\left(2^{\frac{t}{d}}\right) = \ln n_0 + \frac{t}{d}\ln 2$

   $\ln n(t) - \ln n_0 = \frac{t}{d}\ln 2$

   $d\left(\ln n(t) - \ln n_0\right) = t\ln 2$

   $d = \dfrac{t \ln 2}{\ln n(t) - \ln n_0}$

   Substituting $t = 4$, $n_0 = 256{,}000$, and $n(4) = 1{,}440{,}000$

   So, $d = \dfrac{4 \ln 2}{\ln(n(4)) - \ln n_0}$

   $= \dfrac{4 \ln 2}{\ln(1{,}440{,}000) - \ln(115{,}200)} \approx 12.425$

2. Thorium has a half-life of 24.5 days. How long does it take for 80% of a sample to decay?

   This question asks to find the time t when 20% is left, that is, when $\dfrac{m(t)}{m_0} = 0.20$

   $m(t) = m_0 2^{-\frac{t}{h}}$ Solve for t.

   $\dfrac{m(t)}{m_0} = 2^{-\frac{t}{h}}$

   $0.2 = 2^{-\frac{t}{h}}$ Take log or both sides.

   $\ln .2 = \ln\left(2^{-\frac{t}{h}}\right) = -\frac{t}{h}\ln 2$

   So $t = \dfrac{-h \ln .2}{\ln 2}$

   Substituting $h = 24.5$;

   $t = \dfrac{-h \ln .2}{\ln 2} = \dfrac{-(24.5) \ln .2}{\ln 2} \approx 56.89$ days

B. Newton's Law of Cooling states that the rate of cooling of a object is proportional to the temperature difference between the object and its surroundings, provide that the temperature difference is not too large. It can be shown that $T(t) = T_s + D_0 e^{-kt}$ where T_s is the temperature of the surrounding enviroment, D_0 is the initial temperature difference, and k is a positive constant that is associated with the cooling object.

3. A bowl of soup that is 115°F is placed in an air-conditioned room at 70°F. After 5 minutes, the temperature of the soup is 100°F.

   a) What is the temperature after 10 minutes?

   First find for k. When $T(5) = 100°F$, $T_s = 70°F$, and $D_0 = 115 - 70 = 45°F$. Substituting in

   $$T(t) = T_s + D_0 e^{-kt}$$
   $$100 = 70 + 45e^{-5k}$$
   $$30 = 45e^{-5k}$$
   $$\frac{2}{3} = e^{-5k}$$
   $$\ln\left(\frac{2}{3}\right) = -5k$$
   $$k = -\frac{\ln\left(\frac{2}{3}\right)}{5} \approx 0.081$$

   $T(t) = 70 + 45e^{-0.081t}$
   Now find $T(10)$.
   $T(10) = 70 + 45e^{-0.081(10)} \approx 90.01°F$

   b) How long will it take the soup cool to 80°F?

   Find t so that $T(t) = 80$. Substituting into
   $$T(t) = 70 + 45e^{-0.081t}$$
   $$80 = 70 + 45e^{-0.081t}$$
   $$10 = 45e^{-0.081t}$$
   $$\frac{2}{9} = e^{-0.081t}$$
   $$-0.081t = \ln\left(\frac{2}{9}\right)$$
   $$t = \frac{\ln\left(\frac{2}{9}\right)}{-0.081} \approx 19 \text{ minutes}$$

4. Karen placed $6000 into a retirement account paying 9% compounded quarterly. How many years will it take for this account to reach $25,000?

$$A = P\left(1 + \frac{r}{n}\right)^{nt}$$

Using $A = 25000$, $P = 6000$, $r = 0.09$, and $n = 4$, find t.

$$25000 = 6000\left(1 + \frac{.09}{4}\right)^{4t} = 6000(1.0225)^{4t}$$

$$4.166667 = (1.0225)^{4t}$$

$$\ln(4.166667) = \ln(1.0225)^{4t} = 4t \ln(1.0225)$$

$$t = \frac{\ln(4.166667)}{4\ln(1.0225)} \approx 16.03 \text{ years}$$

However, since interest is paid at the end of each quarter, we move up to the next quarter and $t = 16\frac{1}{4}$ years.

C. Logarithms make quantities which vary over very large ranges more manageable. Three commonly used logarithmic scales are:
1. **pH scale**: used to measure the acid and bases.
   pH = – logarithm of the concentration of hydrogen ion measured in moles per liter.
2. **Ricther scale**: used to measure earthquake.
   $m = \log \frac{I}{S} = \log I - \log S$, where I is the intensity of the earthquake and S is the intensity of a "standard" earthquake.
3. **The Decibel scale**: used to measure loudness of sounds.
   $\beta = 10 \log \frac{I}{I_0}$, where I_0 is a reference intensity of $10^{-12} \frac{\text{watts}}{\text{m}^2}$ at a frequency of 100 hertz.

5. The hydrogen ion concentration of a shampoo is 8.2×10^{-8}. Find its pH.

$$\text{pH} = -\log(8.2 \times 10^{-8})$$
$$= -\left(\log(8.2) + \log(10^{-8})\right)$$
$$= -(0.91 - 8) = 7.09$$

6. On October 17, 1989, the Loma Prieta Earthquake rocked the Santa Cruz mountains. Researchers initially measured the earthquake as 6.9 on the Richter scale. Later, the earthquake was re-evaluated as 7.1 on the Richter scale. How do the two measures compare in intensity?

$m = \log I - \log S \Rightarrow m + \log S = \log I$

So, $I = 10^{(m + \log S)} = S\, 10^m$

Comparing the two intensities, we have

$$\frac{S\, 10^{7.1}}{S\, 10^{6.9}} = 10^{7.1-6.9} = 10^{.2} = 1.58$$

So a 7.1 earthquake is 1.58 times more intense than a 6.9 earthquake.

7. A car alarms advertise a sound reading of watts/m². Find the intensity level of the car alarm.

$\beta = 10 \log \dfrac{I}{I_0}$,

Chapter 7

Systems of Equations and Inequalities

Section 7.1   Pairs of Lines
Section 7.2   Systems of Linear Equations
Section 7.3   Inconsistent and Dependent Systems
Section 7.4   Algebra of Matrices
Section 7.5   Inverses of Matrices and Matrix Equations
Section 7.6   Determinants and Cramer's Rule
Section 7.7   Nonlinear Systems
Section 7.8   Systems of Inequalities
Section 7.9   Application: Linear Programming
Section 7.10 Application: Partial Fractions

Section 7.1 Pairs of Lines

Key Ideas
A. Solutions to systems of equations.
B. Steps of the substitution method.
C. Steps of the elimination method.
D. Applications of systems of equation.

A. A **system of equations** is a set of equations with common variables. A **solution** to a system of equations is a point that simultaneously makes each equation of the system a true statement. A system of two equations in two unknowns can have *infinitely many* solutions, *exactly one* solution, or *no* solutions.

1. Determine if the given ordered pair a solution to the system of equations.

   a) (3, 4)
   $$\begin{cases} 2x - y = 2 \\ x + 2y = 11 \end{cases}$$

   Is (3, 4) a solution to $2x - y = 2$?
   Does $2(3) - (4) \stackrel{?}{=} 2$ YES!
   Is (3, 4) a solution to $x + 2y = 11$?
   Does $(3) + 2(4) \stackrel{?}{=} 11$ YES!

   Yes, (3, 4) is a solution to this system.

   b) (−2, 5)
   $$\begin{cases} 3x + y = -1 \\ 8x + 3y = 1 \end{cases}$$

   Is (−2, 5) a solution to $3x + y = -1$?
   Does $3(-2) + (5) \stackrel{?}{=} -1$ YES!
   Is (−2, 5) a solution to $8x + 3y = 1$?
   Does $8(-2) + 3(5) \stackrel{?}{=} 1$ NO!

   No, (−2, 5) is not a solution to the system.

B. Steps of the **substitution method** for two equations in two variables.
   1. Pick one *equation* and one *variable*. Solve the equation for this variable in terms of the other variable.
   2. Substitute this value into the *other equation*. (It is very important that you substitute this value into a *different* equation.)
   3. Solve for the second variable.
   4. Substitute the value found in Step 3 into the first equation and solve for the *first variable*.
   5. Check your solution in *both* equations.

2. Solve the system
$$\begin{cases} 3x + y = -1 \\ -2x + 3y = 8 \end{cases}$$

Solve $3x + y = -1$ for y (*since it is easier*)
$y = -3x - 1$
Substitute for y in the second equation;
$-2x + 3(-3x - 1) = 8$
$-2x + -9x - 3 = 8$
$-11x = 11$
$x = -1$
Substituting this value of x back into $y = -3x - 1$ yields:
$y = -3(-1) - 1 = 3 - 1 = 2$

Check $(-1, 2)$:
Does $3(-1) + (2) \stackrel{?}{=} -1$ YES!
Does $-2(-1) + 3(2) \stackrel{?}{=} 8$ YES!

3. Solve the system
$$\begin{cases} 3x + 2y = 3 \\ x - 2y = 3 \end{cases}$$

Solve $x - 2y = 3$ for x (*since it is easier*)
$x = 2y + 3$
Substitute for x in the first equation;
$3(2y + 3) + 2y = 3$
$6y + 9 + 2y = 3$
$8y = -6$
$y = -\frac{3}{4}$

Substituting this value back into the first equation yields:
$x = 2y + 3 = 2\left(-\frac{3}{4}\right) + 3$
$= -\frac{3}{2} + 3 = \frac{3}{2}$

Check $\left(\frac{3}{2}, -\frac{3}{4}\right)$:
Does $3\left(\frac{3}{2}\right) + 2\left(-\frac{3}{4}\right) \stackrel{?}{=} 3$
$\frac{9}{2} - \frac{3}{2} \stackrel{?}{=} 3$ YES!
Does $\left(\frac{3}{2}\right) - 2\left(-\frac{3}{4}\right) \stackrel{?}{=} 3$
$\frac{3}{2} + \frac{3}{2} \stackrel{?}{=} 3$ YES!

4. Solve the system
$$\begin{cases} 2x - 3y = 9 \\ -4x + 6y = 4 \end{cases}$$

Solve $2x - 3y = 9$ for x.
$2x = 3y + 9$
$x = \dfrac{3y + 9}{2}$

Substitute this value of x into the second equation yields:
$-4\left(\dfrac{3y + 9}{2}\right) + 6y = 4$
$-6y - 18 + 6y = 4$
$-18 = 4$ Impossible!

This system has no solution. Further examination of this system shows that these lines are parallel.

C. Step of the **elimination method** for two equations in two variables.
1. Pick the variable to eliminated.
2. Multiply the first equation by a *nonzero* constant and the second equation by another *nonzero* constant so that the coefficients of the variable (chosen in Step 1) are additive inverses of each other.
3. Add the corresponding sides of the two equations and to create a third equation *(without the variable chosen in Step 1)*.
4. Solve this third equation for the variable not eliminated.
5. Substitute the value found in Step 4 into either equation and solve for the other variable.
6. Check your solution.

5. Solve the system
$$\begin{cases} 3x + 2y = -5 \\ x + 4y = 5 \end{cases}$$

$3x + 2y = -5$
$x + 4y = 5$

Eliminate the x.
$\begin{array}{l} 3x + 2y = -5 \\ -3(x + 4y = 5) \end{array} \Rightarrow \begin{array}{l} 3x + 2y = -5 \\ \underline{-3x - 12y = -15} \\ -10y = -20 \end{array}$

So, $y = 2$. Substituting for y and solve for x
$x + 4(2) = 5 \Rightarrow x + 8 = 5 \Rightarrow x = -3$

OR

Eliminate the y.
$\begin{array}{l} -2(3x + 2y = -5) \\ x + 4y = 5 \end{array} \Rightarrow \begin{array}{l} -6x - 4y = 10 \\ \underline{x + 4y = 5} \\ -5x = 15 \end{array}$

So, $x = -3$. Substituting for x,
$3(-3) + 2y = -5 \Rightarrow -9 + 2y = -5 \Rightarrow 2y = 4$
$y = 2$

By either method the solution is $(-3, 2)$.

6. Solve the system
$$\begin{cases} \frac{3}{4}x + y = 1 \\ 3x + 2y = 1 \end{cases}$$

$\frac{3}{4}x + y = 1$
$3x + 2y = 1$
Eliminate the x.
$-4(\frac{3}{4}x + y = 1) \implies \begin{array}{r} -3x - 4y = -4 \\ 3x + 2y = 1 \\ \hline -2y = -3 \end{array}$
$3x + 2y = 1$

So, $y = \frac{3}{2}$. Substituting for y and solving for x.

$3x + 2\left(\frac{3}{2}\right) = 1 \implies 3x + 3 = 1 \implies 3x = -2 \implies x = -\frac{2}{3}$.

OR

Eliminate the y.

$-2(\frac{3}{4}x + y = 1) \implies \begin{array}{r} -\frac{3}{2}x - 2y = -2 \\ 3x + 2y = 1 \\ \hline \frac{3}{2}x = -1 \end{array}$
$3x + 2y = 1$

So, $x = -\frac{2}{3}$. Substituting for x and solve for y.

$3\left(-\frac{2}{3}\right)x + 2y = 1 \implies -2 + 2y = 1 \implies 2y = 3$
$\implies y = \frac{3}{2}$

By either method the solution is $\left(-\frac{2}{3}, \frac{3}{2}\right)$.

D. Systems of linear equations frequently appear in word problems and many students have problems with word problems. Remember to:
1. Read through the entire problem before starting.
2. Assign letters to denote the variable quantities in the problem and *define* what each letter represents. This helps you (and whoever reads your work) to understand what value you are looking for.
3. Translate the given information from English sentences into equivalent mathematical equations involving the variables.

7. The sum of two numbers is 28 and their difference is 11. Find the two numbers.

Let x = the larger number.
Let y = the other number.
Then "the sum of two numbers is 28" translates to:
$x + y = 28$. And "their difference is 11" translates to: $x - y = 11$

$\begin{array}{r} x + y = 28 \\ x - y = 11 \\ \hline 2x = 37 \end{array}$ ADD.

So $x = \frac{37}{2}$. Substitute into the first equation yields:

$\frac{37}{2} + y = 28 \qquad\qquad y = 28 - \frac{37}{2} = \frac{19}{2}$

Solution: the numbers are $\frac{37}{2}$ and $\frac{19}{2}$

8. A movie theater charges $3.50 for children and $5.50 for adults. On a certain day, 625 people saw a movie and the theater collected $2787.50 in receipts. How many children and how many adults were admitted?

Let x = the number of children admitted.
Let y = the number of adults admitted.

$x + y = 625$ 625 people attended.
$3.50x + 5.50y = 2787.50$ Receipt total 2787.50.

Clear the decimals in the second equation.
$x + y = 625$
$10(3.50x + 5.50y) = 10(2787.50)$

Eliminate the x.
$x + y = 625$
$35x + 55y = 27875$

$\begin{array}{l} -35(x + y = 625) \\ 35x + 55y = 27875 \end{array} \Rightarrow \begin{array}{l} -35x - 35y = -21875 \\ \underline{35x + 55y = 27875} \\ 20y = 6000 \end{array}$

So $y = 300$ and $x + (300) = 625 \Rightarrow x = 325$

Solution: 325 children, 300 adults.

9. A city plans to plant 50 new trees in a subdivision. If birch trees cost $30 each and oak tree are $40 each, how many of each kind of trees can they plant for $1650?

Let x = the number of birch trees.
Let y = the number of each trees.

$x + y = 50$ 50 trees to plant.
$30x + 40y = 1650$ Cost equation.

Multiply the first equation by –30 and add to second equation.

$\begin{array}{l} -30(x + y = 50) \\ 30x + 40y = 1650 \end{array} \Rightarrow \begin{array}{l} -30x - 30y = -1500 \\ \underline{30x + 40y = 1650} \\ 10y = 150 \end{array}$

So $y = 15$ and $x + (15) = 50 \Rightarrow x = 35$

Solution: 35 birch trees and 15 oak trees.

Section 7.2 Systems of Linear Equations

Key Ideas
A. Linear equations and equivalent systems.
B. Gaussian elimination and triangular form.
C. Gaussian elimination using matrices.

A. A **linear equation in n variables** is an equation of the form $a_1x_1 + a_2x_2 + a_3x_3 + \cdots + a_nx_n = c$ where $a_1, a_2, a_3, \ldots, a_n$, and c are constants and $x_1, x_2, x_3, \ldots,$ and x_n are variables. When $n \leq 4$, the letters $x, y, z,$ and w, are used in place of x_1, x_2, x_3, x_4. Each term in a linear equation is either a constant or a constant multiple of *one* of the variable to the first power. Two systems of linear equations are **equivalent** if they have the same solution set.

1. Determine if the equation is linear or not.

   a) $\sqrt{21}x + 3z = 5$

   Yes, $\sqrt{21}$, 3 and 5 are all constants.

   b) $-2xy + z - 7w = 3$

   NO. The first term, $-2xy$, is not linear.

   c) $\frac{1}{3}x - 4y + 51z = -11$

   Yes, $\frac{1}{3}$, -4, 51, and -11 are all constants.

   d) $x + 2\sqrt{y} - 3z = 23$

   NO, in the second term, $2\sqrt{y}$, y is not to the first power.

   e) $\frac{5}{x} + 2y - 13z + 7w = 0$

   NO, in the first term, $\frac{5}{x}$, x is not to the first power (x is to the -1 power).

B. Gaussian elimination is a systematic procedure for transforming a system of linear equations to an equivalent system (that is easier to solve). The only allowable algebraic operations for transformations of one system into another system are called **elementary row operations** and are:
1. Add a multiple of one equation to another equation.

2. Multiply an equation by a *nonzero* constant.
3. Rearrange the order of the equations in the set of equations.

The **Gaussian Elimination Procedure** is:
1. Eliminate x in two equations by adding a multiple of one equation to another equation. Use two *different* pairs of equations to form these new equations.
2. Eliminate the y from the two equations formed in Step 1. The system is now in **triangular form**.
3. Solve for z.
4. Substitute for z in one of the equations formed in Step 1 and solve for y.
5. Substitute for z and y in an original equation and solve for x.
6. Check the solution found in *all three equations*.

2. Solve the system using Gaussian elimination
$$\begin{cases} x + 3y - z = 4 \\ 2x + 5y + z = -1 \\ 3x - y + 2z = -3 \end{cases}$$

$$\begin{cases} x + 3y - z = 4 \\ 2x + 5y + z = -1 \\ 3x - y + 2z = -3 \end{cases}$$

Eliminate the x from the second and third equations using the first equation.

$$\begin{array}{l} -2(x + 3y - z = 4) \\ 2x + 5y + z = -1 \end{array} \Rightarrow \begin{array}{l} -2x - 6y + 2z = -8 \\ \underline{2x + 5y + z = -1} \\ -y + 3z = -9 \end{array}$$

$$\begin{array}{l} -3(x + 3y - z = 4) \\ 3x - y + 2z = -3 \end{array} \Rightarrow \begin{array}{l} -3x - 9y + 3z = -12 \\ \underline{3x - y + 2z = -3} \\ -10y + 5z = -15 \end{array}$$

$$\begin{cases} x + 3y - z = 4 \\ -y + 3z = -9 \\ -10y + 5z = -15 \end{cases} \text{ or } \begin{cases} x + 3y - z = 4 \\ -y + 3z = -9 \\ 2y - z = 3 \end{cases}$$

Dividing the third equation by -5, to simplify it. Now eliminate y in the third equation, using the second equation.

$$\begin{array}{l} 2(-y + 3z = -9) \\ 2y - z = 3 \end{array} \Rightarrow \begin{array}{l} -2y + 6z = -18 \\ \underline{2y - z = 3} \\ 5z = -15 \end{array}$$

System in triangular form:
$$\begin{cases} x + 3y - z = 4 \\ -y + 3z = -9 \\ 5z = -15 \end{cases}$$

Solve the last equation for z.

$5z = -15 \quad \Rightarrow \quad z = -3$

Substitute back into equation 2 and solve for y.

$-y + 3(-3) = -9 \Rightarrow -y - 9 = -9 \Rightarrow -y = 0 \Rightarrow y = 0$

Substitute back into equation 1 and solve for x.

$x + 3(0) - (-3) = 4 \Rightarrow x + 3 = 4 \Rightarrow x = 1$

Check:
$(1) + 3(0) - (-3) = 4 \quad$ yes
$2(1) + 5(0) + (-3) = -1 \quad$ yes
$3(1) - (0) + 2(-3) = -3 \quad$ yes

Solution $(1, 0, -3)$.

3. Solve the system using Gaussian elimination
$$\begin{cases} x + y + 3z = 1 \\ 3x + 2y + 3z = 1 \\ 2x + y + 2z = -3 \end{cases}$$

$$\begin{cases} x + y + 3z = 1 \\ 3x + 2y + 3z = 1 \\ 2x + y + 2z = -3 \end{cases}$$

Eliminate the x from the second and third equations using the first equation.

$$\begin{array}{l} 3(x + y + 3z = 1) \\ -(3x + 2y + 3z = 1) \end{array} \Rightarrow \begin{array}{l} 3x + 3y + 9z = 3 \\ \underline{-3x - 2y - 3z = -1} \\ y + 6z = 2 \end{array}$$

$$\begin{array}{l} -2(x + y + 3z = 1) \\ 2x + y + 2z = -3 \end{array} \Rightarrow \begin{array}{l} -2x - 2y - 6z = -2 \\ \underline{2x + y + 2z = -3} \\ -y - 4z = -5 \end{array}$$

$$\begin{cases} x + y + 3z = 1 \\ y + 6z = 2 \\ -y - 4z = -5 \end{cases}$$

Now eliminate y in the third equation, using the *new* second equation.

$$\begin{array}{l} y + 6z = 2 \\ \underline{-y - 4z = -5} \\ 2z = -3 \end{array}$$

System in triangular form: $\begin{cases} x + y + 3z = 1 \\ y + 6z = 2 \\ 2z = -3 \end{cases}$

Solve the last equation for z.

$$2z = -3 \Rightarrow z = -\frac{3}{2}$$

Substitute for y and solve for y.

$$y + 6\left(-\frac{3}{2}\right) = 2 \Rightarrow y - 9 = 2 \Rightarrow y = 11$$

Substitute for y and z and solve for x.

$$x + (11) + 3\left(-\frac{3}{2}\right) = 1 \Rightarrow x + (11) + 3\left(-\frac{3}{2}\right) = 1$$

$$x + \frac{13}{2} = 1 \Rightarrow x = -\frac{11}{2}$$

Check:

$$\left(-\frac{11}{2}\right) + (11) + 3\left(-\frac{3}{2}\right) = -\frac{11}{2} + 11 - \frac{9}{2}$$
$$= -10 + 11 = 1 \quad \text{Yes}$$

$$3\left(-\frac{11}{2}\right) + 2(11) + 3\left(-\frac{3}{2}\right) = -\frac{33}{2} + 22 - \frac{9}{2}$$
$$= -21 + 22 = 1 \quad \text{Yes}$$

$$2\left(-\frac{11}{2}\right) + (11) + 2\left(-\frac{3}{2}\right) = -11 + 11 - 3 = -3 \quad \text{Yes}$$

Solution: $\left(-\frac{11}{2}, 11, -\frac{3}{2}\right)$

C. Instead of writing the equations in a system out in full, we write only the coefficients and constants in a rectangular array, called a **matrix**. Each equation corresponds to a **row** in the matrix and each variable to a **column** of the matrix The constants also contribute to a column. One matrix is transformed to an equivalent matrix by the row operations given as **elementary row operations** in part B. There is no *one* way to solve a matrix, so it is helpful to yourself and others (who read your work) to leave a trail pf the row operations you perform. The are several systematic procedure to simplify matrices, Gaussian Elimination is one procedure.

4. Write the system in matrix form and solve.
$$\begin{cases} x + 2y + 3z = 1 \\ 2x - y - 3z = 1 \\ x + 3y + 4z = -2 \end{cases}$$

$$\begin{bmatrix} 1 & 2 & 3 & 1 \\ 2 & -1 & -3 & 1 \\ 1 & 3 & 4 & -2 \end{bmatrix}$$

$\rightarrow \begin{array}{c} R_2 \rightarrow R_2 - 2R_1 \\ R_3 \rightarrow R_3 - R_1 \end{array}$ $\begin{bmatrix} 1 & 2 & 3 & 1 \\ 0 & -5 & -9 & -1 \\ 0 & 1 & 1 & -3 \end{bmatrix}$

$\rightarrow \begin{array}{c} R_2 \rightarrow R_3 \\ R_3 \rightarrow R_2 \end{array}$ $\begin{bmatrix} 1 & 2 & 3 & 1 \\ 0 & 1 & 1 & -3 \\ 0 & -5 & -9 & -1 \end{bmatrix}$

$\rightarrow$ $R_3 \rightarrow R_3 + 5R_2$ $\begin{bmatrix} 1 & 2 & 3 & 1 \\ 0 & 1 & 1 & -3 \\ 0 & 0 & -4 & -16 \end{bmatrix}$

$\rightarrow$ $R_3 \rightarrow -\frac{1}{4}R_3$ $\begin{bmatrix} 1 & 2 & 3 & 1 \\ 0 & 1 & 1 & -3 \\ 0 & 0 & 1 & 4 \end{bmatrix}$

So $z = 4$, Substituting back,

$y + z = -3 \Rightarrow y + (4) = -3 \Rightarrow \underline{y = -7}$

$x + 2y + 3z = 1 \Rightarrow x + 2(-7) + 3(4) = 1$

$x - 2 = 1 \Rightarrow \underline{x = 3}$

Solution is the point $(3, -7, 4)$.

Check the solution to make sure it satisfies all three equations in the original system.

5. Write the system in matrix form and solve.
$$\begin{cases} x + y + 5z = 2 \\ x - 2y + w = 1 \\ 2x + 10z + w = 0 \\ -x + 2y - 5z - 4w = -1 \end{cases}$$

$$\begin{bmatrix} 1 & 1 & 5 & 0 & 2 \\ 1 & -2 & 0 & 1 & 1 \\ 2 & 0 & 10 & 1 & 0 \\ -1 & 2 & -5 & -4 & -1 \end{bmatrix}$$ Don't forget to insert 0 for the coefficents of missing terms.

$$\xrightarrow{\begin{array}{l} R_2 \to R_2 - R_1 \\ R_3 \to R_3 - 2R_1 \\ R_4 \to R_4 + R_1 \end{array}} \begin{bmatrix} 1 & 1 & 5 & 0 & 2 \\ 0 & -3 & -5 & 1 & -1 \\ 0 & -2 & 0 & 1 & -4 \\ 0 & 3 & 0 & -4 & 1 \end{bmatrix}$$

It is often easier to use an entry that is 1 to eliminate the terms below it in a column. In this case we would like to create a 1 in the second row, second column. The easiest way to do this is to add row 3 and row 4 and put the result into row 2. (This also gives a 0 in the row 2, column 3 position). The old row 2 must be moved to make row. Do not throw away row 2!

$$\xrightarrow{\begin{array}{l} R_2 \to R_4 + R_3 \\ R_3 \to R_2 \\ R_4 \to R_3 \end{array}} \begin{bmatrix} 1 & 1 & 5 & 0 & 2 \\ 0 & 1 & 0 & -3 & -3 \\ 0 & 3 & 5 & -1 & 1 \\ 0 & -2 & 0 & 1 & -4 \end{bmatrix}$$

$$\xrightarrow{\begin{array}{l} R_3 \to R_3 - 2R_2 \\ R_4 \to R_4 + 2R_2 \end{array}} \begin{bmatrix} 1 & 1 & 5 & 0 & 2 \\ 0 & 1 & 0 & -3 & -3 \\ 0 & 0 & 5 & 8 & 10 \\ 0 & 0 & 0 & -5 & -10 \end{bmatrix}$$

Solve for w.

$-5w = -10, \Rightarrow \underline{w = 2}$

Substituting back and solve for z.

$5z + 8w = 10 \Rightarrow 5z + 8(2) = 10 \Rightarrow 5z = -6$

$\Rightarrow \underline{z = -\dfrac{6}{5}}$

$y - 3w = -3 \Rightarrow y - 3(2) = -3 \Rightarrow \underline{y = 3}$

$x + y + 5z = 2 \Rightarrow x + (3) + 5\left(-\dfrac{6}{5}\right) = 2$

$x + 3 - 6 = 2 \Rightarrow \underline{x = 5}$

Solution is the point $\left(5, 3, -\dfrac{6}{5}, 2\right)$.

Check the solution to make sure it satisfies all four equations of the original system.

Section 7.3 Inconsistent and Dependent Systems

Key Ideas
A. Inconsistent systems.
B. Dependent systems.

A. A system that has no solution is called an **inconsistent system**. If we use Gaussian elimination to try to change a system to triangular form, and if one of the equations is false, then the system is inconsistent. The false equation always has the form $0 = N$, where N is nonzero number.

B. A system that has infinitely many solution is called an **dependent system**. The complete solution to such a system will have one or more variables that are arbitrary, and the other variables will depend on only the arbitrary one(s). If we use Gaussian elimination to try to change a system into triangular form, discarding equations of the form $0 = 0$, and if the resulting system has more variables than equations then the system is dependent.

Find the complete solution to each system of equations, or show that none exists.

1. $\begin{cases} x - 7y + 5z = 6 \\ 2x + 3y - z = 5 \\ 4x - 11y + 9z = 10 \end{cases}$

$\begin{bmatrix} 1 & -7 & 5 & 6 \\ 2 & 3 & -1 & 5 \\ 4 & -11 & 9 & 10 \end{bmatrix}$

$\rightarrow \begin{array}{c} R_2 \rightarrow R_2 - 2R_1 \\ R_3 \rightarrow R_3 - 4R_1 \end{array}$ $\begin{bmatrix} 1 & -7 & 5 & 6 \\ 0 & 17 & -11 & -7 \\ 0 & 17 & -11 & -14 \end{bmatrix}$

$\rightarrow \atop R_3 \rightarrow R_3 - R_1$ $\begin{bmatrix} 1 & -7 & 5 & 6 \\ 0 & 17 & -11 & -7 \\ 0 & 0 & 0 & -7 \end{bmatrix}$

Since the last row corresponds to the equation $0 = -7$, which is false, there is No solution.

2. Write the system in matrix form and solve.
$$\begin{cases} x - 2y = 9 \\ x + 3y - 5z = -6 \\ 2x - y - 3z = 9 \end{cases}$$

$$\begin{bmatrix} 1 & -2 & 0 & 9 \\ 1 & 3 & -5 & -6 \\ 2 & -1 & -3 & 9 \end{bmatrix}$$

$\rightarrow$ $R_2 \rightarrow R_2 - R_1$
$R_3 \rightarrow R_3 - 2R_1$
$$\begin{bmatrix} 1 & -2 & 0 & 9 \\ 0 & 5 & -5 & -15 \\ 0 & 3 & -3 & -9 \end{bmatrix}$$

$\rightarrow$ $R_2 \rightarrow \frac{1}{5}R_2$
$R_3 \rightarrow \frac{1}{5}R_3$
$$\begin{bmatrix} 1 & -2 & 0 & 9 \\ 0 & 1 & -1 & -3 \\ 0 & 1 & -1 & -3 \end{bmatrix}$$

$\rightarrow$
$R_3 \rightarrow R_3 - 2R_1$
$$\begin{bmatrix} 1 & -2 & 0 & 9 \\ 0 & 1 & -1 & -3 \\ 0 & 0 & 0 & 0 \end{bmatrix}$$

Deleting the last row, the resulting matrix is in triangular form and represents 2 equation in 3 unknowns, so there is no unique solution.

$$\begin{bmatrix} 1 & -2 & 0 & 9 \\ 0 & 1 & -1 & -3 \end{bmatrix}$$

Solve each equation in terms of z.

$y - z = -3 \Rightarrow \underline{y = -3 + z}$

$x - 2y = 9 \Rightarrow x - 2(-3 + z) = 9 \Rightarrow x + 6 - 2z = 9 \Rightarrow$
$\underline{x = 3 - 2z}$

Solution:
$x = 3 - 2z$
$y = -3 + z$
$z =$ any real number

3. Write the system in matrix form and solve.
$$\begin{cases} x - y - z = -3 \\ 2x + 3y + 4z = 3 \\ 2x - y + z = 3 \end{cases}$$

$$\begin{bmatrix} 1 & -1 & -1 & -3 \\ 2 & 3 & 4 & 3 \\ 2 & -1 & 1 & 3 \end{bmatrix}$$

$\rightarrow \begin{matrix} R_2 \rightarrow R_2 - 2R_1 \\ R_3 \rightarrow R_3 - 2R_1 \end{matrix}$ $\begin{bmatrix} 1 & -1 & -1 & -3 \\ 0 & 5 & 6 & 9 \\ 0 & 1 & 3 & 9 \end{bmatrix}$

$\rightarrow \begin{matrix} R_2 \rightarrow R_3 \\ R_3 \rightarrow R_2 \end{matrix}$ $\begin{bmatrix} 1 & -1 & -1 & -3 \\ 0 & 1 & 3 & 9 \\ 0 & 5 & 6 & 9 \end{bmatrix}$

$\rightarrow$ $R_3 \rightarrow R_3 - 5R_2$ $\begin{bmatrix} 1 & -1 & -1 & -3 \\ 0 & 1 & 3 & 9 \\ 0 & 0 & -9 & -36 \end{bmatrix}$

$\rightarrow$ $R_3 \rightarrow -\frac{1}{9}R_3$ $\begin{bmatrix} 1 & -1 & -1 & -3 \\ 0 & 1 & 3 & 9 \\ 0 & 0 & 1 & 4 \end{bmatrix}$

So $z = 4$. Substituting back,

$y + 3z = 9 \Rightarrow y + 3(4) = 9 \Rightarrow \underline{y = -3}$

$x - y - z = -3 \Rightarrow x - (-3) - (4) = -3 \Rightarrow \underline{x = -2}$

Solution is the point $(-2, -3, 4)$.

Check the solution to make sure it satisfies all three equations of the original system.

4. Write the system in matrix form and solve.
$$\begin{cases} x - 2z + w = 2 \\ x + y + 5z + w = 7 \\ 2x + y + 3z + 2w = 9 \\ -3x - y + z - 3w = -13 \end{cases}$$

$$\begin{bmatrix} 1 & 0 & -2 & 1 & 2 \\ 1 & 1 & 5 & 1 & 7 \\ 2 & 1 & 3 & 2 & 9 \\ -3 & -1 & 1 & -3 & -13 \end{bmatrix}$$

$\rightarrow \begin{matrix} R_2 \rightarrow R_2 - R_1 \\ R_3 \rightarrow R_3 - 2R_1 \\ R_4 \rightarrow R_4 + 3R_1 \end{matrix}$ $\begin{bmatrix} 1 & 0 & -2 & 1 & 2 \\ 0 & 1 & 7 & 0 & 5 \\ 0 & 1 & 7 & 0 & 5 \\ 0 & 1 & 7 & 0 & 7 \end{bmatrix}$

$\rightarrow \begin{matrix} R_3 \rightarrow R_3 - R_2 \\ R_4 \rightarrow R_4 - R_2 \end{matrix}$ $\begin{bmatrix} 1 & 0 & -2 & 1 & 2 \\ 0 & 1 & 7 & 0 & 5 \\ 0 & 0 & 0 & 0 & 0 \\ 0 & 0 & 0 & 0 & 2 \end{bmatrix}$

Since the last equation, $0 = 2$, is false, there is NO solution.

5. Write the system in matrix form and solve.
$$\begin{cases} x - 2y - 2z + 2w = -3 \\ x + 2y + 2z + 4w = 5 \\ x - 6y - 6z = -11 \\ 2x - 2y - 2z + 5w = -2 \end{cases}$$

$$\begin{bmatrix} 1 & -2 & -2 & 2 & -3 \\ 1 & 2 & 2 & 4 & 5 \\ 1 & -6 & -6 & 0 & -11 \\ 2 & -2 & -2 & 5 & -2 \end{bmatrix}$$

$\rightarrow \begin{array}{c} R_2 \rightarrow R_2 - R_1 \\ R_3 \rightarrow R_3 - R_1 \\ R_4 \rightarrow R_4 - 2R_1 \end{array}$
$\begin{bmatrix} 1 & -2 & -2 & 2 & -3 \\ 0 & 4 & 4 & 2 & 8 \\ 0 & -4 & -4 & -2 & -8 \\ 0 & 2 & 2 & 1 & 4 \end{bmatrix}$

$\rightarrow \begin{array}{c} R_2 \rightarrow R_4 \\ \\ R_4 \rightarrow R_2 \end{array}$
$\begin{bmatrix} 1 & -2 & -2 & 2 & -3 \\ 0 & 2 & 2 & 1 & 4 \\ 0 & -4 & -4 & -2 & -8 \\ 0 & 4 & 4 & 2 & 8 \end{bmatrix}$

$\rightarrow \begin{array}{c} R_3 \rightarrow R_3 + 2R_2 \\ R_4 \rightarrow R_4 - 2R_2 \end{array}$
$\begin{bmatrix} 1 & -2 & -2 & 2 & -3 \\ 0 & 2 & 2 & 1 & 4 \\ 0 & 0 & 0 & 0 & 0 \\ 0 & 0 & 0 & 0 & 0 \end{bmatrix}$

Since the last two rows are all zeroes, we can delete them from the matrix.

$\rightarrow R_2 \rightarrow \frac{1}{2}R_2$
$\begin{bmatrix} 1 & -2 & -2 & 2 & -3 \\ 0 & 1 & 1 & \frac{1}{2} & 2 \end{bmatrix}$

Since this matrix has 2 equations in 4 variables we need to solve for y in terms of z and w.

$y + z + \frac{1}{2}w = 2 \implies \underline{y = 2 - z - \frac{1}{2}w}$

Substituting for y:
$x - 2y - 2z + 2w = -3$
$x - 2\left(2 - z - \frac{1}{2}w\right) - 2z + 2w = -3 \implies$
$x - 4 + 2z + w - 2z + 2w = -3 \implies x - 4 + 3w = -3$
So, $\underline{x = 1 - 3w}$

Solution:
$x = 1 - 3w$
$y = 2 - z - \frac{1}{2}w$
z = any real number
w = any real number

6. Write the system in matrix form and solve.
$$\begin{cases} x + 2y - w = 0 \\ x + y + z + w = 9 \\ 3x + 4y - z + w = -5 \\ 2x + 3y - z = -6 \end{cases}$$

$$\begin{bmatrix} 1 & 2 & 0 & -1 & 0 \\ 1 & 1 & 1 & 1 & 8 \\ 3 & 4 & -1 & 1 & -1 \\ 2 & 3 & -1 & 0 & -6 \end{bmatrix}$$

$\rightarrow$ $\begin{array}{l} R_2 \rightarrow R_2 - R_1 \\ R_3 \rightarrow R_3 - 3R_1 \\ R_4 \rightarrow R_4 - 2R_1 \end{array}$ $\begin{bmatrix} 1 & 2 & 0 & -1 & 0 \\ 0 & -1 & 1 & 2 & 8 \\ 0 & -2 & -1 & 4 & -5 \\ 0 & -1 & -1 & 2 & -6 \end{bmatrix}$

$\rightarrow$ $R_2 \rightarrow -R_2$ $\begin{bmatrix} 1 & 2 & 0 & -1 & 0 \\ 0 & 1 & -1 & -2 & -8 \\ 0 & -2 & -1 & 4 & -5 \\ 0 & -1 & -1 & 2 & -6 \end{bmatrix}$

$\rightarrow$ $\begin{array}{l} R_3 \rightarrow R_3 + 2R_2 \\ R_4 \rightarrow R_4 + R_2 \end{array}$ $\begin{bmatrix} 1 & 2 & 0 & -1 & 0 \\ 0 & 1 & -1 & -2 & -8 \\ 0 & 0 & -3 & 0 & -21 \\ 0 & 0 & -2 & 0 & -14 \end{bmatrix}$

Again, since it is easier to work with a 1, we create a 1 in row 3, column 3 position.

$\rightarrow$ $R_3 \rightarrow -\frac{1}{3}R_3$ $\begin{bmatrix} 1 & 2 & 0 & -1 & 0 \\ 0 & 1 & -1 & -2 & -8 \\ 0 & 0 & 1 & 0 & 7 \\ 0 & 0 & -2 & 0 & -14 \end{bmatrix}$

$\rightarrow$ $R_4 \rightarrow R_4 + 2R_3$ $\begin{bmatrix} 1 & 2 & 0 & -1 & 2 \\ 0 & 1 & -1 & -2 & -8 \\ 0 & 0 & 1 & 0 & 7 \\ 0 & 0 & 0 & 0 & 0 \end{bmatrix}$

Delete the last row since it all 0.
$$\begin{bmatrix} 1 & 2 & 0 & -1 & 2 \\ 0 & 1 & -1 & -2 & -8 \\ 0 & 0 & 1 & 0 & 7 \end{bmatrix}$$

Since this matrix has 3 equations in 4 variables, the system is dependent. Solve the last row for z.

$\underline{z = 7}$

Substitute for z.

$y - z - 2w = -8$

$y - (7) - 2w = -8 \implies y - 2w = -1$

Solve one for y in terms of w: $\quad \underline{y = -1 + 2w}$

Solve for x in terms of w: $\quad x + 2y - w = 2$
$x + 2(-1 + 2w) - w = 2 \implies x - 2 + 4w - w = 2$
$x - 2 + 3w = 2 \implies \underline{x = 4 - 3w}$

Solution:
$x = 4 - 3w \qquad y = -1 + 2w$
$z = 7 \qquad\qquad w = $ any real number

Section 7.4 Algebra of Matrices

Key Ideas
A. Matrix terminology, addition/subtraction and scalar multiplication.
B. Inner product and matrix multiplication.

A. The **dimension** of a matrix is the number of rows by the number of columns, expressed as $R \times C$ and read as 'R by C'. The numbers in a matrix are called **entries** and are referred to by their row, column location. So in matrix A, the term $a_{i,j}$ is the i-row j-column entry. Two matrices are equal if and only if they have the same dimensions and the corresponding entries are equal. Two matrices can be be added or subtracted whenever they are the same dimension. The resulting matrix is found by adding (or subtracting) the corresponding entries. A constant times a matrix is found by multiplying every entry in the matrix by this constant. The constant is called **scalar** and the product is called the **scalar product**.

Carry out the algebraic operation indicated in each exercises, or explain why it cannot be performed.

1. $\begin{bmatrix} 2 & -5 \\ 0 & 7 \end{bmatrix} + \begin{bmatrix} 8 & -5 \\ -5 & 4 \end{bmatrix}$

$\begin{bmatrix} 2 & -5 \\ 0 & 7 \end{bmatrix} + \begin{bmatrix} 8 & -5 \\ -5 & 4 \end{bmatrix}$ Add the corresponding terms

$= \begin{bmatrix} 2+8 & -5+(-5) \\ 0+(-5) & 7+4 \end{bmatrix}$

$= \begin{bmatrix} 10 & -10 \\ -5 & 11 \end{bmatrix}$

2. $\begin{bmatrix} 2 & 1 & 8 \\ 11 & -3 & 0 \\ 7 & 0 & 9 \\ -2 & 1 & 8 \end{bmatrix} - \begin{bmatrix} 8 & -2 & -1 \\ 0 & -5 & 4 \\ 1 & 3 & -2 \\ 1 & 0 & 3 \end{bmatrix}$

$\begin{bmatrix} 2 & 1 & 8 \\ 11 & -3 & 0 \\ 7 & 0 & 9 \\ -2 & 1 & 8 \end{bmatrix} - \begin{bmatrix} 8 & -2 & -1 \\ 0 & -5 & 4 \\ 1 & 3 & -2 \\ 1 & 0 & 3 \end{bmatrix}$

$= \begin{bmatrix} 2-8 & 1-(-2) & 8-(-1) \\ 11-0 & -3-(-5) & 0-4 \\ 7-1 & 0-3 & 9-(-2) \\ -2-1 & 1-0 & 8-3 \end{bmatrix}$

$= \begin{bmatrix} 10 & 3 & 9 \\ 11 & 2 & -4 \\ 6 & -3 & 11 \\ -3 & 1 & 5 \end{bmatrix}$

3. $2\begin{bmatrix} 4 & 0 & -2 & 3 & 4 \\ -3 & 2 & -6 & 0 & -5 \\ 6 & 2 & -5 & 4 & 1 \\ 3 & -2 & 1 & 0 & 3 \end{bmatrix}$

$2\begin{bmatrix} 4 & 0 & -2 & 3 & 4 \\ -3 & 2 & -6 & 0 & -5 \\ 6 & 2 & -5 & 4 & 1 \\ 3 & -2 & 1 & 0 & 3 \end{bmatrix}$

$= \begin{bmatrix} 2\cdot 4 & 2\cdot 0 & 2\cdot -2 & 2\cdot 3 & 2\cdot 4 \\ 2\cdot -3 & 2\cdot 2 & 2\cdot -6 & 2\cdot 0 & 2\cdot -5 \\ 2\cdot 6 & 2\cdot 2 & 2\cdot -5 & 2\cdot 4 & 2\cdot 1 \\ 2\cdot 3 & 2\cdot -2 & 2\cdot 1 & 2\cdot 0 & 2\cdot 3 \end{bmatrix}$

$= \begin{bmatrix} 8 & 0 & -4 & 6 & 8 \\ -6 & 4 & -12 & 0 & -10 \\ 12 & 4 & -10 & 8 & 2 \\ 6 & -4 & 2 & 0 & 6 \end{bmatrix}$

4. $2\begin{bmatrix} 4 & 0 & 7 \\ 3 & 4 & -11 \\ 2 & -6 & 0 \end{bmatrix} - \begin{bmatrix} 6 & 2 \\ -5 & 4 \\ 1 & 3 \\ -2 & 1 \\ 0 & 3 \end{bmatrix}$

Although the scalar product, $2\begin{bmatrix} 4 & 0 & 7 \\ 3 & 4 & -11 \\ 2 & -6 & 0 \end{bmatrix}$ is possible. the result is a 3×3 matrix and that cannot be added to a 5×2 matrix.

B. If A is a **row matrix** (dimension $1 \times n$) and B is a **column matrix** (dimension $n \times 1$) then the **inner product**, AB is the sum of the product to the corresponding entries. That is:

$$[a_1 \; a_2 \; \cdots \; a_n] \cdot \begin{bmatrix} b_1 \\ b_2 \\ \vdots \\ b_n \end{bmatrix} = [a_1 b_1 + a_2 b_2 + \ldots + a_n b_n]$$

Suppose that A is an $m \times n$ matrix and B is an $n \times k$ matrix, then the **product** $C = AB$ of two matrices is an $m \times k$ matrix, where the entry $c_{i,j}$ is the inner product of the ith row of A and the jth column of B. Helpful hints when finding matrix products.
1. Always check the dimensions first, to make sure that the product is possible and to find the dimension of the resulting matrix.
2. The location of each entry in the resulting matrix tells how to generate it.

The matrix product is associative but is *not* commutative, that is,
$(AB)C = A(BC)$ but $AB \neq BA$.

Find each matrix product or explain why it cannot be performed.

5. $\begin{bmatrix} -3 & 1 \\ -4 & 5 \\ -1 & 2 \end{bmatrix} \cdot \begin{bmatrix} -7 & 2 \\ 3 & 1 \end{bmatrix}$

This is the product of a 3×2 times a 2×2 matrix. The results is possible and is a 3×2.

$$\begin{bmatrix} -3 & 1 \\ -4 & 5 \\ -1 & 2 \end{bmatrix} \cdot \begin{bmatrix} -7 & 2 \\ 3 & 1 \end{bmatrix}$$

$$= \begin{bmatrix} -3(-7) + 1(3) & -3(2) + 1(1) \\ -4(-7) + 5(3) & -4(2) + 5(1) \\ -1(-7) + 2(3) & -1(2) + 2(1) \end{bmatrix}$$

$$= \begin{bmatrix} 21 + 3 & -6 + 1 \\ 28 + 15 & -8 + 5 \\ 7 + 6 & -2 + 2 \end{bmatrix}$$

$$= \begin{bmatrix} 24 & -5 \\ 43 & -3 \\ 7 & 0 \end{bmatrix}$$

6. $\begin{bmatrix} 2 & 1 & 0 & 4 & 2 \\ 0 & 0 & -3 & -1 & 2 \end{bmatrix} \cdot \begin{bmatrix} -7 & 2 \\ 2 & 5 \\ 0 & -2 \\ 4 & 3 \end{bmatrix}$

This is the product of a 2×5 times a 4×2 matrix. Since the 5 and 4 are not equal this product is not possible.

7. $\begin{bmatrix} 0 & 9 & -3 & 2 \\ 1 & 0 & 3 & -4 \\ 5 & 0 & -1 & 2 \end{bmatrix} \cdot \begin{bmatrix} 1 & 0 \\ 2 & 1 \\ 3 & 0 \\ 0 & 5 \end{bmatrix}$

This is the product of a 3×4 times a 4×2 matrix. The results is possible and is a 3×2.

$\begin{bmatrix} 0 & 9 & -3 & 2 \\ 1 & 0 & 3 & -4 \\ 5 & 0 & -1 & 2 \end{bmatrix} \cdot \begin{bmatrix} 1 & 0 \\ 2 & 1 \\ 3 & 0 \\ 0 & 5 \end{bmatrix} =$

$\begin{bmatrix} 0(1) + 9(2) - 3(3) + 2(0) & 0(0) + 9(1) - 3(0) + 2(5) \\ 1(1) + 0(2) + 3(3) - 4(0) & 1(0) + 0(1) + 3(0) - 4(5) \\ 5(1) + 0(2) - 1(3) + 2(0) & 5(0) + 0(1) - 1(0) + 2(5) \end{bmatrix}$

$= \begin{bmatrix} 0 + 18 - 9 + 0 & 0 + 9 - 0 + 10 \\ 1 + 0 + 9 - 0 & 0 + 0 + 0 - 20 \\ 5 + 0 - 3 + 0 & 0 + 0 - 0 + 10 \end{bmatrix}$

$= \begin{bmatrix} 9 & 19 \\ 10 & -20 \\ 2 & 10 \end{bmatrix}$

8. $\begin{bmatrix} 4 & 0 & 7 \\ 3 & 4 & -11 \\ 2 & -6 & 0 \end{bmatrix} \cdot \begin{bmatrix} 6 & 2 & -5 \\ 4 & 1 & 3 \\ -2 & 1 & 3 \end{bmatrix}$

This is the product of a 3×3 times a 3×3 matrix. The results is possible and is a 3×3.

$\begin{bmatrix} 4 & 0 & 7 \\ 3 & 4 & -11 \\ 2 & -6 & 0 \end{bmatrix} \cdot \begin{bmatrix} 6 & 2 & -5 \\ 4 & 1 & 3 \\ -2 & 1 & 3 \end{bmatrix}$

$= \begin{bmatrix} 24 + 0 - 14 & 8 + 0 + 7 & -20 + 0 + 21 \\ 18 + 16 + 22 & 6 + 4 - 11 & -15 + 12 - 33 \\ 12 - 24 + 0 & 4 - 6 + 0 & -10 - 18 + 0 \end{bmatrix}$

$= \begin{bmatrix} 10 & 15 & 1 \\ 56 & -1 & -36 \\ -12 & -2 & -28 \end{bmatrix}$

Write the system of equations as a matrix equation.

9. $\begin{cases} x + 2y - z = 0 \\ x + y + z = 9 \\ 3x + 4y - z = -5 \end{cases}$

$\begin{bmatrix} 1 & 2 & -1 \\ 1 & 1 & 1 \\ 3 & 4 & 1 \end{bmatrix} \cdot \begin{bmatrix} x \\ y \\ z \end{bmatrix} = \begin{bmatrix} 0 \\ 9 \\ -5 \end{bmatrix}$

10. $\begin{cases} x - 3y + z + w = -3 \\ x + 4y + z + 4w = 5 \\ x - 6y - 6z = -11 \\ x - 2y - 2z + 5w = 3 \end{cases}$

$\begin{bmatrix} 1 & -3 & 1 & 1 \\ 1 & 4 & 1 & 4 \\ 1 & -6 & -6 & 0 \\ 1 & -2 & -2 & 5 \end{bmatrix} \cdot \begin{bmatrix} x \\ y \\ z \\ w \end{bmatrix} = \begin{bmatrix} -3 \\ 5 \\ -11 \\ 3 \end{bmatrix}$

11. Let $A = \begin{bmatrix} 4 & 3 & 3 \\ -1 & 0 & -1 \\ -4 & -4 & -3 \end{bmatrix}$.

   a) Find A^2 and A^3.

$A^2 = \begin{bmatrix} 4 & 3 & 3 \\ -1 & 0 & -1 \\ -4 & -4 & -3 \end{bmatrix} \cdot \begin{bmatrix} 4 & 3 & 3 \\ -1 & 0 & -1 \\ -4 & -4 & -3 \end{bmatrix}$

$= \begin{bmatrix} 1 & 0 & 0 \\ 0 & 1 & 0 \\ 0 & 0 & 1 \end{bmatrix}$

$A^3 = A \cdot A^2$

$= \begin{bmatrix} 4 & 3 & 3 \\ -1 & 0 & -1 \\ -4 & -4 & -3 \end{bmatrix} \cdot \begin{bmatrix} 1 & 0 & 0 \\ 0 & 1 & 0 \\ 0 & 0 & 1 \end{bmatrix}$

$= \begin{bmatrix} 4 & 3 & 3 \\ -1 & 0 & -1 \\ -4 & -4 & -3 \end{bmatrix}.$

   b) Find a general formula for A^n.

Since $A^3 = A$

$A^{odd} = \begin{bmatrix} 4 & 3 & 3 \\ -1 & 0 & -1 \\ -4 & -4 & -3 \end{bmatrix}$

$A^{even} = \begin{bmatrix} 1 & 0 & 0 \\ 0 & 1 & 0 \\ 0 & 0 & 1 \end{bmatrix}$

Section 7.5 Inverses of Matrices and Matrix Equations

Key Ideas
A. The identity matrix and the inverses of 2 by 2 matrices.
B. Determinants and inverses of larger matrices.
C. Matrix equations.

A. The main diagonal of a square $n \times n$ matrix is the diagonal of a matrix whose entries are $a_{1,1}\ a_{2,2}\ a_{3,3}\ \ldots\ a_{nn}$. The **identity matrix** I_n is the $n \times n$ matrix where each entry of the main diagonal is a 1, and all other entries are 0. Identity matrices behave like the number 1 in the sense that $A \cdot I_n = A$ and $I_n \cdot B = B$, whenever these products are defined. If A and B are $n \times n$ matrices where $A \cdot B = B \cdot A = I_n$, then we say that B is the **inverse** of A, and we write $B = A^{-1}$. The concept of the inverse of a matrix is analogous to that of a reciprocal of a real number; however, *not all square matrices have inverses.* When A is a 2×2 matrix, the inverse of A can be calculated (if it exists) in the following formula. If $A = \begin{bmatrix} a & b \\ c & d \end{bmatrix}$, then $A^{-1} = \frac{1}{ad-bc}\begin{bmatrix} d & -b \\ -c & a \end{bmatrix}$, provided $ad - bc \neq 0$. The quantity, $ad - bc$ is called the **determinant** of the matrix A. Matrix A has a inverse if and only if the determinant of A does not equal 0.

1. Are the following pairs of matrices inverses of each other?

   This question asks; Does $AB = I_n$? Check by finding the product.

   a) $\begin{bmatrix} 4 & 2 \\ 3 & 3 \end{bmatrix}$ and $\begin{bmatrix} \frac{1}{2} & -\frac{1}{3} \\ -\frac{1}{2} & \frac{2}{3} \end{bmatrix}$.

   Does $\begin{bmatrix} 4 & 2 \\ 3 & 3 \end{bmatrix} \cdot \begin{bmatrix} \frac{1}{2} & -\frac{1}{3} \\ -\frac{1}{2} & \frac{2}{3} \end{bmatrix} = I_2$?

   $\begin{bmatrix} 4 & 2 \\ 3 & 3 \end{bmatrix} \cdot \begin{bmatrix} \frac{1}{2} & -\frac{1}{3} \\ -\frac{1}{2} & \frac{2}{3} \end{bmatrix} = \begin{bmatrix} 1 & 0 \\ 0 & 1 \end{bmatrix}$

   So, these two matrices are inverses of each other.

   b) $\begin{bmatrix} 1 & 1 & -1 \\ 1 & 2 & -1 \\ -1 & -1 & 2 \end{bmatrix}$ and $\begin{bmatrix} 3 & -1 & 1 \\ -1 & 1 & 0 \\ 1 & 0 & 1 \end{bmatrix}$.

   Does $\begin{bmatrix} 1 & 1 & -1 \\ 1 & 2 & -1 \\ -1 & -1 & 2 \end{bmatrix} \cdot \begin{bmatrix} 3 & -1 & 1 \\ -1 & 1 & 0 \\ 1 & 0 & 1 \end{bmatrix} = I_3$?

   $\begin{bmatrix} 1 & 1 & -1 \\ 1 & 2 & -1 \\ -1 & -1 & 2 \end{bmatrix} \cdot \begin{bmatrix} 3 & -1 & 1 \\ -1 & 1 & 0 \\ 1 & 0 & 1 \end{bmatrix} = \begin{bmatrix} 1 & 0 & 0 \\ 0 & 1 & 0 \\ 0 & 0 & 1 \end{bmatrix}$

   So, these two matrices are inverses of each other.

2. Find the inverse of each 2 × 2 matrix, if it exists.

a) $\begin{bmatrix} 4 & 7 \\ 3 & 5 \end{bmatrix}$

Using the formula: $A = \begin{bmatrix} a & b \\ c & d \end{bmatrix}$

$\rightarrow A^{-1} = \dfrac{1}{ad-bc}\begin{bmatrix} d & -b \\ -c & a \end{bmatrix}$

First find $ad - bc$. $\quad 4 \cdot 5 - 7 \cdot 3 = -1$

So $A^{-1} = -\begin{bmatrix} 5 & -7 \\ -3 & 4 \end{bmatrix}$

$= \begin{bmatrix} -5 & 7 \\ 3 & -4 \end{bmatrix}$

b) $\begin{bmatrix} 9 & 12 \\ 6 & 8 \end{bmatrix}$

Using the same formula as in Question 2a above.
first find $ad - bc$. $\quad 9 \cdot 8 - 6 \cdot 12 = 72 - 72 = 0$
So this matrix has no inverse.

c) $\begin{bmatrix} 2 & 3 \\ -3 & 5 \end{bmatrix}$

Using the same formula as in Question 2a above.
first find $ad - bc$. $\quad 2 \cdot 5 - 3 \cdot (-3) = 10 + 9 = 19$

So $A^{-1} = \dfrac{1}{ad-bc}\begin{bmatrix} d & -b \\ -c & a \end{bmatrix}$

$A^{-1} = \dfrac{1}{19}\begin{bmatrix} 5 & -3 \\ 3 & 2 \end{bmatrix}$

$= \begin{bmatrix} \dfrac{5}{19} & -\dfrac{3}{19} \\ \dfrac{3}{19} & \dfrac{2}{19} \end{bmatrix}$

B. One technique to calculate the inverse of a $n \times n$ matrix A, is to augment it with I_n, the identity matrix. Then use row operations to change the left side to the identity matrix. That is start with $\begin{bmatrix} A & I_n \end{bmatrix}$ and use row operations to transform it to the matrix $\begin{bmatrix} I_n & A^{-1} \end{bmatrix}$.

3. Find the inverse of each matrix.

a) $A = \begin{bmatrix} 1 & 0 & 2 \\ 6 & 1 & 16 \\ 2 & 0 & 3 \end{bmatrix}$

Augment A with I_3

$\begin{bmatrix} 1 & 0 & 2 & 1 & 0 & 0 \\ 6 & 1 & 16 & 0 & 1 & 0 \\ 2 & 0 & 3 & 0 & 0 & 1 \end{bmatrix}$

$\rightarrow \begin{array}{c} R_2 \rightarrow R_2 - 6R_1 \\ R_3 \rightarrow R_3 - 2R_1 \end{array} \begin{bmatrix} 1 & 0 & 2 & 1 & 0 & 0 \\ 0 & 1 & 4 & -6 & 1 & 0 \\ 0 & 0 & -1 & -2 & 0 & 1 \end{bmatrix}$

$\rightarrow \begin{array}{c} \\ \\ R_3 \rightarrow -R_3 \end{array} \begin{bmatrix} 1 & 0 & 2 & 1 & 0 & 0 \\ 0 & 1 & 4 & -6 & 1 & 0 \\ 0 & 0 & 1 & 2 & 0 & -1 \end{bmatrix}$

$\rightarrow \begin{array}{c} R_1 \rightarrow R_1 - 2R_3 \\ R_2 \rightarrow R_2 - 4R_3 \end{array} \begin{bmatrix} 1 & 0 & 0 & -3 & 0 & 2 \\ 0 & 1 & 0 & -14 & 1 & 4 \\ 0 & 0 & 1 & 2 & 0 & -1 \end{bmatrix}$

So $A^{-1} = \begin{bmatrix} -3 & 0 & 2 \\ -14 & 1 & 4 \\ 2 & 0 & -1 \end{bmatrix}$

Check:

$\begin{bmatrix} 1 & 0 & 2 \\ 6 & 1 & 16 \\ 2 & 0 & 3 \end{bmatrix} \cdot \begin{bmatrix} -3 & 0 & 2 \\ -14 & 1 & 4 \\ 2 & 0 & -1 \end{bmatrix} \stackrel{?}{=} \begin{bmatrix} 1 & 0 & 0 \\ 0 & 1 & 0 \\ 0 & 0 & 1 \end{bmatrix}$

b) $B = \begin{bmatrix} 1 & 2 & 1 \\ 1 & 1 & -1 \\ 3 & 4 & -2 \end{bmatrix}$

Augment B with I_3

$\begin{bmatrix} 1 & 2 & 1 & 1 & 0 & 0 \\ 1 & 1 & -1 & 0 & 1 & 0 \\ 3 & 4 & -2 & 0 & 0 & 1 \end{bmatrix}$

$\rightarrow \begin{array}{c} R_2 \rightarrow R_2 - 1R_1 \\ R_3 \rightarrow R_3 - 3R_1 \end{array} \begin{bmatrix} 1 & 2 & 1 & 1 & 0 & 0 \\ 0 & -1 & -2 & -1 & 1 & 0 \\ 0 & -2 & -4 & -3 & 0 & 1 \end{bmatrix}$

$\rightarrow R_2 \rightarrow -R_2 \begin{bmatrix} 1 & 2 & 1 & 1 & 0 & 0 \\ 0 & 1 & 2 & 1 & -1 & 0 \\ 0 & -2 & -4 & -3 & 0 & -1 \end{bmatrix}$

$\rightarrow \begin{array}{c} \\ \\ R_3 \rightarrow R_3 + 2R_2 \end{array} \begin{bmatrix} 1 & 2 & 1 & 1 & 0 & 0 \\ 0 & 1 & 2 & 1 & -1 & 4 \\ 0 & 0 & 0 & -1 & -2 & -1 \end{bmatrix}$

Since the last row of the matrix $\begin{bmatrix} 1 & 2 & 1 \\ 0 & 1 & 2 \\ 0 & 0 & 0 \end{bmatrix}$ we have gotten is all 0, this matrix has no inverse.

c) $C = \begin{bmatrix} 7 & -3 & -3 \\ -1 & 1 & 0 \\ -1 & 0 & 1 \end{bmatrix}$

Augment C with I_3

$$\begin{bmatrix} 7 & -3 & -3 & 1 & 0 & 0 \\ -1 & 1 & 0 & 0 & 1 & 0 \\ -1 & 0 & 1 & 0 & 0 & 1 \end{bmatrix}$$

$\xrightarrow[R_2 \to R_1]{R_1 \to -R_2}$ $\begin{bmatrix} 1 & -1 & 0 & 0 & -1 & 0 \\ 7 & -3 & -3 & 1 & 0 & 0 \\ -1 & 0 & 1 & 0 & 0 & 1 \end{bmatrix}$

$\xrightarrow[R_3 \to R_3 + R_1]{R_2 \to R_2 - 7R_1}$ $\begin{bmatrix} 1 & -1 & 0 & 0 & -1 & 0 \\ 0 & 4 & -3 & 1 & 7 & 0 \\ 0 & -1 & 1 & 0 & -1 & 1 \end{bmatrix}$

$\xrightarrow[R_3 \to R_2]{R_2 \to -R_3}$ $\begin{bmatrix} 1 & -1 & 0 & 0 & -1 & 0 \\ 0 & 1 & -1 & 0 & 1 & -1 \\ 0 & 4 & -3 & 1 & 7 & 0 \end{bmatrix}$

$\xrightarrow[R_3 \to R_3 - 4R_2]{}$ $\begin{bmatrix} 1 & -1 & 0 & 0 & -1 & 0 \\ 0 & 1 & -1 & 0 & 1 & -1 \\ 0 & 0 & 1 & 1 & 3 & 4 \end{bmatrix}$

$\xrightarrow{R_2 \to R_2 + R_3}$ $\begin{bmatrix} 1 & -1 & 0 & 0 & -1 & 0 \\ 0 & 1 & 0 & 1 & 4 & 3 \\ 0 & 0 & 1 & 1 & 3 & 4 \end{bmatrix}$

$\xrightarrow{R_1 \to R_1 + R_2}$ $\begin{bmatrix} 1 & 0 & 0 & 1 & 3 & 3 \\ 0 & 1 & 0 & 1 & 4 & 3 \\ 0 & 0 & 1 & 1 & 3 & 4 \end{bmatrix}$

So $C^{-1} = \begin{bmatrix} 1 & 3 & 3 \\ 1 & 4 & 3 \\ 1 & 3 & 4 \end{bmatrix}$

Check:

$\begin{bmatrix} 7 & -3 & -3 \\ -1 & 1 & 0 \\ -1 & 0 & 1 \end{bmatrix} \cdot \begin{bmatrix} 1 & 3 & 3 \\ 1 & 4 & 3 \\ 1 & 3 & 4 \end{bmatrix} \stackrel{?}{=} \begin{bmatrix} 1 & 0 & 0 \\ 0 & 1 & 0 \\ 0 & 0 & 1 \end{bmatrix}$

C. A system of linear equations can be written as a **matrix equation**, $A \cdot X = B$, where A is the **coefficient matrix**, X is the **unknown matrix**, and B is the **constant matrix**. The solution to the matrix equation is found by multiplying both sides by A^{-1} provided the inverse exists. So the matrix equation $A \cdot X = B$

$$A^{-1}(A \cdot X) = A^{-1}B$$
$$(A^{-1} \cdot A)X = A^{-1}B$$
$$I_n \cdot X = A^{-1}B$$
$$X = A^{-1}B$$

Remember, the matrix product is associative but not is commutative, so the side you multiply by A^{-1} is important.

4. Write the system of equations as a matrix product and use the inverse you found in Problems 1, 2 and 3.

   a) $\begin{cases} 4x + 2y = 7 \\ 3x + 3y = -3 \end{cases}$

   $A \cdot X = B$ can be solve by the matrix product $X = A^{-1}B$ provided A^{-1} exist.

   $$\begin{bmatrix} 4 & 2 \\ 3 & 3 \end{bmatrix} \cdot \begin{bmatrix} x \\ y \end{bmatrix} = \begin{bmatrix} 7 \\ -3 \end{bmatrix}$$

   The inverse was used in Problem 1a.

   $$\begin{bmatrix} x \\ y \end{bmatrix} = \begin{bmatrix} \frac{1}{2} & -\frac{1}{3} \\ -\frac{1}{2} & \frac{2}{3} \end{bmatrix} \cdot \begin{bmatrix} 7 \\ -3 \end{bmatrix}$$

   $$\begin{bmatrix} x \\ y \end{bmatrix} = \begin{bmatrix} \frac{9}{2} \\ -\frac{11}{2} \end{bmatrix}$$

   Check:
   $$4\left(\frac{9}{2}\right) + 2\left(-\frac{11}{2}\right) = 18 - 11 = 7$$
   $$3\left(\frac{9}{2}\right) + 3\left(-\frac{11}{2}\right) = \frac{27}{2} - \frac{33}{2} = -\frac{8}{2} = -4$$

   b) Write the system in matrix form and solve.
   $\begin{cases} 4x + 7y = 2 \\ 3x + 5y = 1 \end{cases}$

   $$\begin{bmatrix} 4 & 7 \\ 3 & 5 \end{bmatrix} \cdot \begin{bmatrix} x \\ y \end{bmatrix} = \begin{bmatrix} 2 \\ 1 \end{bmatrix}$$

   The inverse was found in problem 2a.

   $$\begin{bmatrix} x \\ y \end{bmatrix} = \begin{bmatrix} -5 & 7 \\ 3 & -4 \end{bmatrix} \cdot \begin{bmatrix} 2 \\ 1 \end{bmatrix} = \begin{bmatrix} -3 \\ 2 \end{bmatrix}$$

   Check:
   $4(-3) + 7(2) = -12 + 14 = 2$
   $3(-3) + 5(2) = -9 + 10 = 1$

c) Write the system in matrix form and solve.
$$\begin{cases} 2x + 3y = -95 \\ -3x + 5y = 76 \end{cases}$$

$$\begin{bmatrix} 2 & 3 \\ -3 & 5 \end{bmatrix} \cdot \begin{bmatrix} x \\ y \end{bmatrix} = \begin{bmatrix} -95 \\ 76 \end{bmatrix}$$

The inverse was found in Problem 2c.

$$\begin{bmatrix} x \\ y \end{bmatrix} = \begin{bmatrix} \frac{5}{19} & -\frac{3}{19} \\ \frac{3}{19} & \frac{2}{19} \end{bmatrix} \cdot \begin{bmatrix} -95 \\ 76 \end{bmatrix} = \begin{bmatrix} -37 \\ -7 \end{bmatrix}$$

Check:
2(–37) + 3(–7) = –74 – 21 = –95
–3(–37) + 5(–7) = 111 – 35 = 76

d) Write the system in matrix form and solve.
$$\begin{cases} x + y - z = -3 \\ x + 2y - z = 6 \\ -x - y + 2z = 4 \end{cases}$$

$$\begin{bmatrix} 1 & 1 & -1 \\ 1 & 2 & -1 \\ -1 & -1 & 2 \end{bmatrix} \cdot \begin{bmatrix} x \\ y \\ z \end{bmatrix} = \begin{bmatrix} -3 \\ 6 \\ 4 \end{bmatrix}$$

The inverse was used in Problem 1b.

$$\begin{bmatrix} x \\ y \\ z \end{bmatrix} = \begin{bmatrix} 3 & -1 & 1 \\ -1 & 1 & 0 \\ 1 & 0 & 1 \end{bmatrix} \cdot \begin{bmatrix} -3 \\ 6 \\ 4 \end{bmatrix}$$

So, $\begin{bmatrix} x \\ y \\ z \end{bmatrix} = \begin{bmatrix} -11 \\ 9 \\ 1 \end{bmatrix}$

Check:
(–11) + (9) – (1) = –11 + 9 – 1 = –3
(–11) + 2(9) – (1) = –11 + 18 – 1 = 6
–(–11) – (9) + 2(1) = 11 – 9 + 2 = 4

e) Write the system in matrix form and solve.
$$\begin{cases} x + 2z = 5 \\ 6x + y + 16z = 3 \\ 2x + 3z = -5 \end{cases}$$

$$\begin{bmatrix} 1 & 0 & 2 \\ 6 & 1 & 16 \\ 2 & 0 & 3 \end{bmatrix} \cdot \begin{bmatrix} x \\ y \\ z \end{bmatrix} = \begin{bmatrix} 5 \\ 3 \\ -5 \end{bmatrix}$$

The inverse was found in Problem 3a.

$$\begin{bmatrix} x \\ y \\ z \end{bmatrix} = \begin{bmatrix} -3 & 0 & 2 \\ -14 & 1 & 4 \\ 2 & 0 & -1 \end{bmatrix} \cdot \begin{bmatrix} 5 \\ 3 \\ -5 \end{bmatrix}$$

So, $\begin{bmatrix} x \\ y \\ z \end{bmatrix} = \begin{bmatrix} -25 \\ -87 \\ 15 \end{bmatrix}$

Check:
$(-25) + 2(15) = -25 + 30 = 5$
$6(-25) + (-87) + 16(15) = -150 - 87 + 240 = 3$
$2(-25) + 3(15) = -50 + 45 = 5$

f) Write the system in matrix form and solve.
$$\begin{cases} 7x - 3y - 3z = 4 \\ -x + y = -2 \\ -x + z = 1 \end{cases}$$

$$\begin{bmatrix} 7 & -3 & -3 \\ -1 & 1 & 0 \\ -1 & 0 & 1 \end{bmatrix} \cdot \begin{bmatrix} x \\ y \\ z \end{bmatrix} = \begin{bmatrix} 4 \\ -2 \\ 1 \end{bmatrix}$$

The inverse was found in Problem 2c.

$$\begin{bmatrix} x \\ y \\ z \end{bmatrix} = \begin{bmatrix} 1 & 3 & 3 \\ 1 & 4 & 3 \\ 1 & 3 & 4 \end{bmatrix} \cdot \begin{bmatrix} 4 \\ -2 \\ 1 \end{bmatrix}$$

So, $\begin{bmatrix} x \\ y \\ z \end{bmatrix} = \begin{bmatrix} 1 \\ -1 \\ 2 \end{bmatrix}$

Check:
$7(1) - 3(-1) - 3(2) = 7 + 3 - 6 = 4$
$-(1) + (-1) = -1 - 1 = -2$
$-(1) + (2) = -1 + 2 = 1$

Section 7.6 Determinants and Cramer's Rule

Key Ideas
A. Determinants, minors, and cofactors.
B. Invertibility, or when does matrix A have an inverse?
C. Cramer's rule.

A. The **determinant** of the square matrix A is a value denoted by the symbol $|A|$. For a 1×1 matrix, the determinant is its only entry. When A is the 2×2 matrix $\begin{bmatrix} a & b \\ c & d \end{bmatrix}$ the determinant of A is $|A| = \begin{vmatrix} a & b \\ c & d \end{vmatrix} = ad - bc$. To define the determinant for the general $n \times n$ matrix A, we need some addition terminology. The **minor** $M_{i,j}$ of the element $a_{i,j}$ is the determinant of the matrix obtained by deleting the ith row and jth column of A. The **cofactor** $A_{i,j}$ associated with the element $a_{i,j}$ is: $A_{i,j} = (-1)^{i+j} M_{i,j}$. The determinant of A is obtained by multiplying each element of the first row by its cofactor, and then adding the results. In symbols, this is:

$$|A| = \begin{vmatrix} a_{11} & a_{12} & \cdots & a_{1n} \\ a_{21} & a_{22} & \cdots & a_{2n} \\ \vdots & \vdots & \ddots & \vdots \\ a_{n1} & a_{n2} & \cdots & a_{nn} \end{vmatrix} = a_{11}A_{11} + a_{12}A_{12} + \cdots + a_{1n}A_{1n}$$

This is called *expanding the determinant by the first row*.

Hints:
1. The cofactor of a_{ij} is just the minor of a_{ij} multiplied by either 1 of -1, depending on whether $i + j$ is even or odd.
2. The determinant can be found by expanding about *any* row or column in the matrix; the result will be the same. Usually the row or column with the most zeros is used to expand about..

1. Find the determinant of each matrix.

a) $\begin{bmatrix} 5 & 3 \\ 7 & -2 \end{bmatrix}$

$\begin{vmatrix} 5 & 3 \\ 7 & -2 \end{vmatrix} = 5 \cdot 7 + 3 \cdot (-2) = 35 - 6 = 29$

b) $\begin{bmatrix} 0 & 1 & -2 \\ 4 & 0 & 3 \\ -1 & 2 & 1 \end{bmatrix}$

Expanded about the first row:

$\begin{vmatrix} 0 & 1 & -2 \\ 4 & 0 & 3 \\ -1 & 2 & 1 \end{vmatrix} = 0 \cdot \begin{vmatrix} 0 & 3 \\ 2 & 1 \end{vmatrix} +$

$(-1)(1) \cdot \begin{vmatrix} 4 & 3 \\ -1 & 1 \end{vmatrix} + (1)(-2) \begin{vmatrix} 4 & 0 \\ -1 & 2 \end{vmatrix}$

$= 0 - (4 + 3) - 2(8 - 0) = -7 - 16 = -23$

c) $\begin{bmatrix} 0 & 1 & 0 & 2 \\ 1 & 0 & 1 & 0 \\ 4 & 0 & 3 & 2 \\ -3 & 1 & 2 & 1 \end{bmatrix}$

Expanding about the row,
$$\begin{vmatrix} 0 & 1 & 0 & 2 \\ 1 & 0 & 1 & 0 \\ 4 & 0 & 3 & 2 \\ -3 & 1 & 2 & 1 \end{vmatrix}$$
$$= -1 \begin{vmatrix} 1 & 1 & 0 \\ 4 & 3 & 2 \\ -3 & 2 & 1 \end{vmatrix} - 2 \begin{vmatrix} 1 & 0 & 1 \\ 4 & 0 & 3 \\ -3 & 1 & 2 \end{vmatrix}$$
$$= -1 \cdot (-11) - 2 \cdot 1 = 9$$

Since,
$$\begin{vmatrix} 1 & 1 & 0 \\ 4 & 3 & 2 \\ -3 & 2 & 1 \end{vmatrix} = \begin{vmatrix} 3 & 2 \\ 2 & 1 \end{vmatrix} - 1 \begin{vmatrix} 4 & 2 \\ -3 & 1 \end{vmatrix}$$
$$= (3-4) - (4+6) = -11$$

and,
$$\begin{vmatrix} 1 & 0 & 1 \\ 4 & 0 & 3 \\ -3 & 1 & 2 \end{vmatrix} = \begin{vmatrix} 0 & 3 \\ 1 & 2 \end{vmatrix} + \begin{vmatrix} 4 & 0 \\ -3 & 1 \end{vmatrix}$$
$$= (0-3) + (4-0) = 1$$

B. If a square matrix has a nonzero determinant, then it is called **invertible**, that is, $|A| \neq 0$. A square matrix is *invertible* if and only if A^{-1} exists. Also, if the matrix B is obtained from A by adding a multiple of one row to another, or a multiple of one column to another, then $|A| = |B|$.

2. Find the determinant of $A = \begin{bmatrix} 2 & 1 & 3 \\ 1 & 0 & -4 \\ -2 & 0 & 10 \end{bmatrix}$ and the matrix obtain by replacing the bottom row by $2R_2 + R_3$.

Expanding about the second column:
$$\begin{vmatrix} 2 & 1 & 3 \\ 1 & 0 & -4 \\ -2 & 0 & 10 \end{vmatrix} = -1 \begin{vmatrix} 1 & -4 \\ -2 & 10 \end{vmatrix}$$
$$= -1 \cdot (10 - 8) = -2$$

$[A] \rightarrow \begin{bmatrix} 2 & 1 & 3 \\ 1 & 0 & -4 \\ 0 & 0 & 2 \end{bmatrix}$
$R_3 \rightarrow R_3 + 2R_2$

Expanding $\begin{bmatrix} 2 & 1 & 3 \\ 1 & 0 & -4 \\ 0 & 0 & 2 \end{bmatrix}$ about the last row yields:
$$\begin{vmatrix} 2 & 1 & 3 \\ 1 & 0 & -4 \\ 0 & 0 & 2 \end{vmatrix} = 2 \begin{bmatrix} 2 & 1 \\ 1 & 0 \end{bmatrix} = 2(0-1) = -2$$

C. The solution of linear equations can sometimes be expressed using determinants. A n variable, n equation linear system written in the matrix equation form is:

$$\begin{bmatrix} a_{11} & a_{12} & \cdots & a_{1n} \\ a_{21} & a_{22} & \cdots & a_{2n} \\ \vdots & \vdots & \ddots & \vdots \\ a_{n1} & a_{n2} & \cdots & a_{nn} \end{bmatrix} \begin{bmatrix} x_1 \\ x_2 \\ \vdots \\ x_n \end{bmatrix} = \begin{bmatrix} b_1 \\ b_2 \\ \vdots \\ b_n \end{bmatrix}$$

Let D be the coefficient matrix and let D_{x_i} be the matrix obtained from D by replacing the ith column of D by the numbers $b_1, b_2, ..., b_n$. **Cramer's Rule** states that if $|D| \neq 0$ then:

$$x_1 = \frac{|D_{x_1}|}{|D|}, \quad x_2 = \frac{|D_{x_2}|}{|D|}, \quad \ldots, \quad x_n = \frac{|D_{x_n}|}{|D|}$$

3. Use Cramer's to solve each matrix equation.

$$\begin{bmatrix} 1 & 0 & 2 \\ 2 & 7 & 6 \\ 2 & -1 & 3 \end{bmatrix} \cdot \begin{bmatrix} x \\ y \\ z \end{bmatrix} = \begin{bmatrix} 0 \\ 2 \\ 3 \end{bmatrix}$$

$$D = \begin{bmatrix} 1 & 0 & 2 \\ 2 & 7 & 6 \\ 2 & -1 & 3 \end{bmatrix} \qquad D_x = \begin{bmatrix} 0 & 0 & 2 \\ 2 & 7 & 6 \\ 3 & -1 & 3 \end{bmatrix}$$

$$D_y = \begin{bmatrix} 1 & 0 & 2 \\ 2 & 2 & 6 \\ 2 & 3 & 3 \end{bmatrix} \qquad D_z = \begin{bmatrix} 1 & 0 & 0 \\ 2 & 7 & 2 \\ 2 & -1 & 3 \end{bmatrix}$$

Expanding about row 1:

$$|D| = \begin{vmatrix} 7 & 6 \\ -1 & 3 \end{vmatrix} + 2 \begin{vmatrix} 2 & 7 \\ 2 & -1 \end{vmatrix}$$

$$= (21 + 6) + 2(-2 - 14) = 27 - 32 = -5$$

Expanding about row 1:

$$|D_x| = 2 \begin{vmatrix} 2 & 7 \\ 3 & -1 \end{vmatrix}$$

$$= 2 \cdot (-2 - 21) = -46$$

Expanding about the row 1:

$$|D_y| = \begin{vmatrix} 2 & 6 \\ 3 & 3 \end{vmatrix} + 2 \begin{vmatrix} 2 & 2 \\ 2 & 3 \end{vmatrix}$$

$$= (6 - 18) + 2(6 - 4) = -12 + 4 = -8$$

Expanding about row 1:

$$|D_z| = \begin{vmatrix} 7 & 2 \\ -1 & 3 \end{vmatrix}$$

$$= (21 + 2) = 23$$

$$x = \frac{|D_x|}{|D|} = \frac{-46}{-5} = \frac{46}{5}$$

$$y = \frac{|D_y|}{|D|} = \frac{-8}{-5} = \frac{8}{5}$$

$$z = \frac{|D_z|}{|D|} = \frac{23}{-5} = -\frac{23}{5}$$

Section 7.7   Nonlinear Systems

Key Ideas
A. Algebraic approaches to solving nonlinear systems.
B. Using graphing devices to solve nonlinear systems.

A. Algebraic approaches to solving nonlinear systems can be grouped into three categories. These approaches can be mixed.
   1. **Substitution.** Try this method when an equation can be solved for one variable in terms of the other variables. Then substituted for that variable in the other equations.
   2. **Elimination.** This is a variation of the method used earlier in systems of linear equations. Use this method when a variable can be eliminated by the addition of a multiple of one equation to another equation.
   3. **Factoring.** Use this method when an equation can be set equal to zero *and* factored. Set each factor equal to zero and solve. These results can then be used in the other equations.

   Remember to check each solution in each equation of the original system of equations.

1. Find all real solutions (x, y) of the system of equations.
$$\begin{cases} x^2 - xy - y^2 = 1 \\ 2x + y = 1 \end{cases}$$

Solve the second equation for y and substitute this value into the first equation.

$y = 1 - 2x$

$x^2 - x(1 - 2x) - (1 - 2x)^2 = 1$

$x^2 - x + 2x^2 - (1 - 4x + 4x^2) = 1$

$3x^2 - x - 1 + 4x - 4x^2 = -x^2 + 3x - 1 = 1$

$-x^2 + 3x - 2 = 0 \quad \rightarrow \quad x^2 - 3x + 2 = 0$

$(x - 2)(x - 1) = 0$

$x = 2 \text{ or } x = 1$

Substituting these values for x back into the second equation we get:

when $x = 2$, $y = 1 - 2(2) = -3$
when $x = 1$, $y = 1 - 2(1) = -1$
So the solutions are $(2, -3)$ and $(1, -1)$

Check:
(2, –3)
$(2)^2 - (2)(-3) - (-3)^2 = 4 + 6 - 9 = 1$
$2(2) + (-3) = 4 - 3 = 1 \quad$ Yes.

(1, –1)
$(1)^2 - (1)(-1) - (-1)^2 = 1 + 1 - 1 = 1$
$2(1) + (-1) = 2 - 1 = 1 \quad$ Yes.

2. Find all real solutions (x, y) of the system of equations.
$$\begin{cases} x^2y + xy^2 = 0 \\ x^2 + 3xy + y^2 = 9 \end{cases}$$

Since the first equation can be factored:
$x^2y + xy^2 = xy(x + y) = 0$
So $\underline{x = 0}$ or $\underline{y = 0}$ or $x + y = 0 \rightarrow \underline{x = -y}$
Now substituting each of the possibilities into the second equation;
If $\underline{x = 0}$, then
$(0)^2 + 3(0)y + y^2 = 9 \Rightarrow y^2 = 9 \Rightarrow \underline{y = \pm 3}$
Here we get the solutions $(0, -3)$ and $(0, 3)$.
If $\underline{y = 0}$, then
$x^2 + 3x(0) + (0)^2 = 9 \Rightarrow x^2 = 9 \Rightarrow \underline{x = \pm 3}$
And we get the solutions $(-3, 0)$ and $(3, 0)$.
If $\underline{x = -y}$, then
$(-y)^2 + 3(-y)y + y^2 = 9 \Rightarrow -y^2 = 9$ which is impossible, so no solution.
Check each solution to make sure.
There the only possible solutions are:
$(0, -3), (0, 3), (-3, 0),$ and $(3, 0)$

3. Find all real solutions (x, y) of the system of equations.
$$\begin{cases} x^2 + y^2 + 2y = 10 \\ x^2 + 4y = 7 \end{cases}$$

Since x^2 can be eliminated by subtracting the second equation from the first equation, we eliminate the x
$x^2 + y^2 + 2y = 10$
$\underline{-(x^2 + 4y = 7)}$
$y^2 - 2y = 3 \quad \Rightarrow \quad y^2 - 2y - 3 = 0$
$(y - 3)(y + 1) = 0$
$y - 3 = 0 \quad \Rightarrow \quad \underline{y = 3}$
$y + 1 = 0 \quad \Rightarrow \quad \underline{y = -1}$
Using $\underline{y = 3}$ in the second equation gives:
$x^2 + 4(3) = 7 \quad \Rightarrow \quad x^2 = -5$ Not possible with real numbers.

Using $\underline{y = -1}$ in the second equation gives:
$x^2 + 4(-1) = 7 \quad \Rightarrow \quad x^2 = 11$
$x = \pm \sqrt{11}$

Check each solution to make sure.
There the only possible solutions are:
$\left(-1, -\sqrt{11}\right)$ and $\left(-1, \sqrt{11}\right)$

Note: Another approach is to solve the second equation for y and substituted that value into the first equation. However, this would result with a 4th degree polynomial to factor.

4. Find all real solutions (x, y) of the system of equations.
$$\begin{cases} x^2 + xy - z = 8 \\ x^2 - xy + xz - yz = 0 \\ 2y^2 + 2x + z = -2 \end{cases}$$

Since the second equation can be set equal to zero and factor, we start there:
$$x^2 - xy + xz - yz = 0$$
$$(x - y)(x + z) = 0$$
$x - y = 0 \quad \Rightarrow \quad x = y$ or
$x + z = 0 \quad \Rightarrow \quad x = -z$

Using $\underline{x = y}$ and substituting into the other two equations yield the system
$$\begin{cases} 2y^2 - z = 8 \\ 2y^2 + 2y + z = -2 \end{cases}$$
Adding the two equation eliminates the z:
$4y^2 + 2y = 6 \quad \Rightarrow \quad 4y^2 + 2y - 6 = 0$
$2(2y^2 + y - 3) = 2(2y + 3)(y - 1) = 0$ Factoring.
$2y + 3 = 0 \quad \Rightarrow \quad 2y = -3 \quad \Rightarrow \quad \underline{y = -\frac{3}{2}}$
$y - 1 = 0 \quad \Rightarrow \quad \underline{y = 1}$

$y = -\frac{3}{2}$ Substitute into $2\left(-\frac{3}{2}\right)^2 - z = 8 \quad \underline{z = -\frac{7}{2}}$.
Since $x = y$, this yields the solution: $\left(-\frac{3}{2}, -\frac{3}{2}, -\frac{7}{2}\right)$

$\underline{y = 1}$. substituted into $(1)^2 - z = 8 \quad \Rightarrow \quad \underline{z = -7}$
Since $x = y$, this yields the solution: $(1, 1, -7)$

Using $\underline{x = -z}$ and substituting into the other two equations yield the system.
$$\begin{cases} z^2 - yz - z = 8 \\ 2y^2 - z = -2 \end{cases}$$
Solve the second equation for z: $z = 2y^2 + 2$
and substituting into the first equation yields:
$\left(2y^2 + 2\right)^2 - y\left(2y^2 + 2\right) - \left(2y^2 + 2\right) = 8$
$4y^4 - 2y^3 + 6y^2 - 2y - 6 = 0$
or $2y^4 - y^3 + 3y^2 - y - 3 = 0$
Using synthetic division yields $\underline{y = 1}$ and
$2y^3 + y^2 + 4y + 3 = 0$ which has no other rational root. Using a graphing calculator will yield the only other solution: $\underline{y = -0.7}$
When $y = 1$, $z = 4$ and $x = -4$.
When $y = -0.7$, $z = 2.98$ and $x = -2.98$.

Solutions are:
$\left(-\frac{3}{2}, -\frac{3}{2}, -\frac{7}{2}\right)$,
$(1, 1, -7)$, $(-4, 1, 4)$, and
$(-2.98, -0.7, 2.98)$

[−2, 2] by [−2, 2]

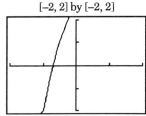

5. Find all real solutions (x, y) of the system of equations.
$$\begin{cases} x - 2y^2 = 6 \\ y - z^2 = 0 \\ x - y^2 + z^2 = 8 \end{cases}$$

Since the first equation can be solved for x in terms of y and the second can be solve for y in terms of z, we substitute this into first to get x in terms of z.
$x = 2y^2 + 6$
$y = z^2$ So $x = 2(z^2)^2 + 6 = 2z^4 + 6$
Substituting into $x - y^2 + z^2 = 8$ yields:
$(2z^4 + 6) - (z^2)^2 + z^2 = 8$
$z^4 + z^2 - 2 = 0$
$(z^2 + 2)(z^2 - 1) = 0$
$z^2 + 2 = 0 \Rightarrow z^2 = -2$ No real solution.
$z^2 - 1 = 0 \Rightarrow z^2 = 1 \Rightarrow \underline{z = \pm 1}$

$\underline{z = 1}$ yields
$x = 2(1)^4 + 6 = 8$ and $y = (1)^2 = 1$ solution: $(8, 1, 1)$

$\underline{z = -1}$ yields
$x = 2(-1)^4 + 6 = 8$ and $y = (-1)^2 = 1$
solution: $(8, 1, -1)$

Solutions are: $(8, 1, 1)$ and $(8, 1, -1)$.

B. Graphing devices are sometimes useful in solving systems of equations in two variables. However, with most graphing devices, an equation must first be expressed in terms of one or more functions of the form $y = f(x)$ before the calculator can graph it. Not all equations can be expressed in this way, so not all systems can be solved this way.

6. Find all real solutions to the system of equation, correct to two decimal places.
$$\begin{cases} x^2y + 3xy + 4y - 4 = 0 \\ x^2 + 4x - y = 6 \end{cases}$$

$x^2y + 3xy + 4y - 4 = 0$ Write y as a function of x.
$x^2y + 3xy + 4y = 4$ Group terms with y to one side.
$y(x^2 + 3x + 4) = 4$ Factor.
$y = \dfrac{4}{x^2 + 3x + 4}$

$x^2 + 4x - y = 6$ Write y as a function of x.
$y = x^2 + 4x - 6$
Solution: $(-5.18, 0.13)$ and $(1.23, 0.43)$
$[-8, 5]$ by $[-2, 2]$

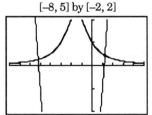

7. Find all real solutions to the system of equation, correct to two decimal places.
$$\begin{cases} e^x y = x^2 \\ (x+2)^2 + (y-2)^2 = 9 \end{cases}$$

Write each equation in y as a function of x form.
$e^x y = x^2$

$(x+2)^2 + (y-2)^2 = 9$

$(y-2)^2 = 9 - (x+2)^2$

$y - 2 = \pm\sqrt{9 - (x+2)^2}$ Take square root of y

$y = 2 \pm \sqrt{9 - (x+2)^2}$ both side, don't forget the $\pm$.

Graph both half of the circle.

Top half: $y = 2 + \sqrt{9 - (x+2)^2}$

Bottom half: $y = 2 - \sqrt{9 - (x+2)^2}$

Solution: $(-1.21, 4.90)$ and $(0.29, 0.062)$

[−3, 1.5] by [−2, 5]

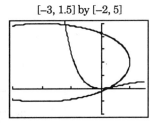

8. Find all real solutions to the system of equation, correct to three decimal places.
$$\begin{cases} xy = x^2 - 3x + 2 \\ x = 3y \end{cases}$$

$xy = x^2 - 3x + 2$ Divide by x and write y as a

$y = \dfrac{x^2 - 3x + 2}{x}$ function of x.

$y = x - 3 + \dfrac{2}{x}$

$x = 3y$

$y = \dfrac{x}{3}$

Solution: $(0.813, 0.271)$ and $(3.688, 1.228)$

[−0.5, 4] by [−1, 3]

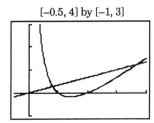

193

Section 7.8 Systems of Inequalities

Key Ideas
A. Graphing inequalities in two variables.
B. Systems of inequalities.

A. The graph of an inequality, in general, consists of a region in the plane whose boundary is the graph of the equation obtained by replacing the inequality sign by an equal sign. The points on the boundary are included when the inequality sign is $\geq$ or $\leq$ and this is indicated by a solid boundary. When the inequality sign is $>$ or $<$, the points on the boundary are excluded and this is indicated by a broken curves. To determine which region of the plane gives the solution set of the inequality, pick a point *not on the curve*, called a **test point**, and test to see if it satisfies the inequality. If it does, then every point in the same region will satisfy the inequality. If it does not, then every point in the same region will not, but every but in the other region will.

Graph the following inequalities.

1. $x + y^2 \geq 4$

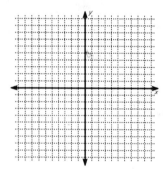

$x + y^2 = 4 \Rightarrow x - 4 = -y^2$ is a parabola, with a vertex at (4, 0) opening to the left.
Using the (0, 0) as a test point:

Test pt.   Test
(0, 0) $(0) + (0)^2 \stackrel{?}{\geq} 4$
         No

Since (0, 0) is inside the region, the solution lies on the outside of the parabola. Because $\geq$ is used, graph the parabola with a solid line.

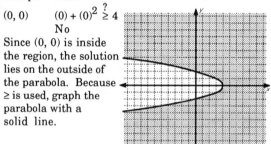

2. $4x^2 + 9y^2 < 36$

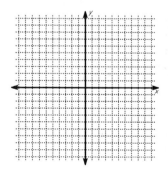

$4x^2 + 9y^2 = 36 \Rightarrow \dfrac{x^2}{9} + \dfrac{y^2}{4} = 1$ is an ellipse.
Using the (0, 0) as a test point:

Test pt.   Test
(0, 0) $4(0)^2 + 9(0)^2 \stackrel{?}{<} 36$
         Yes

Since (0, 0) is in the inside region, the solution lies on the inside. Because $<$ is used, graph the ellipse with a dotted line.

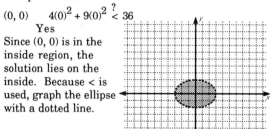

3. $xy > 4$

$xy = 4 \Rightarrow y = \dfrac{4}{x}$ is a hyperbola that has been rotated.
Using the $(-5, -5)$, $(0, 0)$, and $(5, 5)$ as test points:

Test pt.	Test		
$(-5, -5)$	$(-5)(-5) \stackrel{?}{>} 0$		Yes
$(0, 0)$	$(0)(0) \stackrel{?}{>} 0$		NO
$(5, 5)$	$(5)(5) \stackrel{?}{>} 0$		Yes

Here we took more test points than we needed to be on the safe side. Regions and on the "inside of the hyperbolas.

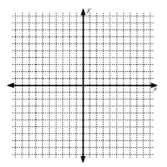
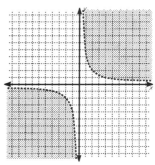

B. The solution to a system of inequalities consist of the intersection of the graphs of each inequality in the system. The **vertices** of a solution set are point where two (or more) inequalities intersect. Vertices may, or may not be part of the solution, but they show where the corners of the solution set lie. A inequality is **linear** if the equality is the equation of a straight line, that is, if it is one of the forms
$$ax + by \geq c \qquad ax + by \leq c \qquad ax + by > c \qquad ax + by < c$$

4. Graph the solution set of the system of inequalities.
$$\begin{cases} x + y < 3 \\ 3x + 2y \geq 5 \end{cases}$$

Graph $x + y = 3$ with a dotted line and graph $3x + 2y = 5$ with a solid line. The vertex here is at $(-1, 4)$.
Test point $(0, 0)$.

Test pt.	Test		
$(0, 0)$	$(0) + (0) \stackrel{?}{<} 3$		Yes
$(0, 0)$	$3(0) + 2(0) \stackrel{?}{\geq} 5$		No

So $(0, 0)$ satisfies the first inequality but not the second.

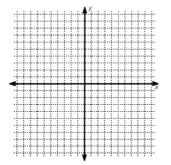
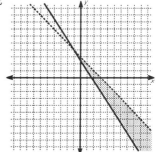

5. Graph the solution set of the system of inequalities.
$$\begin{cases} x^2 + 4x + y \le -6 \\ x + y > -6 \end{cases}$$

Graph $x^2 + 4x + y \le -6$ with a solid line and graph $x + y > -6$ with a dotted line. The vertices are at $(-3, -3)$ and $(0, -6)$
Test point $(0, 0)$.

Test pt.	Test	
$(0, 0)$	$(0)^2 + 4(0) + (0) \stackrel{?}{\le} -6$	No
$(0, 0)$	$(0) + (0) \stackrel{?}{>} -6$	Yes

So $(0, 0)$ satisfies the second inequality but not the first.

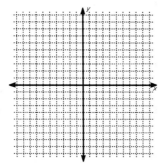

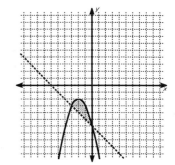

6. Graph the solution set of the system of inequalities.
$$\begin{cases} 3x + y > 9 \\ 2x + y \ge 8 \\ -x + 3y \ge 3 \end{cases}$$

Graph $2x + y \ge 8$ and $-x + 3y \ge 3$ with a solid line and graph $3x + y > 9$ with a dotted line. The vertices are at $(1, 6)$ and $(3, 2)$
Test point $(0, 0)$.

Test pt.	Test	
$(0, 0)$	$3(0) + (0) \stackrel{?}{>} 9$	No
$(0, 0)$	$2(0) + (0) \stackrel{?}{\ge} 8$	No
$(0, 0)$	$-(0) + 3(0) \stackrel{?}{\ge} 3$	No

So $(0, 0)$ satisfies none of the inequalities so the solution does not contain the origin.

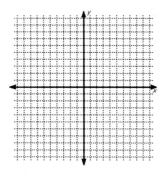

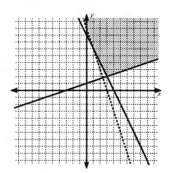

Section 7.9 Application: Linear Programming

Key Ideas
A. Linear programming steps.

A. Linear programming is a mathematical technique used to determine the optimal allocation of resources in many fields. When only two variables are involved, the technique is a step by step graphical procedure. These steps are:
1. Determine the linear inequalities, called **restrictions**.
2. Determine the **objective function**. This is the equation to optimize, either maximize or minimize.
3. Graph the **feasible region**. The feasible region is the region determined by the restrictions obtained in Step 1. Label the lines so that it is easier to determine the vertices.
4. Find the **vertices** of the region, these are the points where two inequalities intersect and are *also* part of the feasible region.
5. Test each vertex in the objective function and pick the vertex (or two vertices) that optimizes the objective function. When there are two vertices that optimize the objective function, then any point on the common boundary will also optimize the objective function.

1. Graph the feasible region, determine the coordinates of its vertices, and the find the maximum value of the objective function on the feasible region by checking its values at the vertices.

$$\begin{cases} x + 3y \leq 42 \\ x + y \leq 16 \\ x \leq 12 \\ x \geq 0, \ y \geq 0 \end{cases}$$

Objective function: $P = 8x + 6y$

Graph the region and find the vertices.

$x + 3y \leq 42$ $x + y \leq 16$
$x \leq 12$ $x \geq 0, y \geq 0$

$x + 3y = 42$ and $x = 0$ intersect at $(0, 14)$
$x + 3y = 42$ and $x + y = 16$ intersect at $(3, 13)$
$x + y = 16$ and $x = 12$ intersect at $(12, 4)$
$x = 12$ and $y = 0$ at $(12, 0)$

Vertices	Objective Function: $P = 8x + 6y$	
(0, 0)	$8(0) + 6(0) = 0$	
(0, 14)	$8(0) + 6(14) = 84$	
(3, 13)	$8(3) + 6(13) = 102$	
(12, 4)	$8(12) + 6(4) = 120$	Maximum
(12, 0)	$8(12) + 6(0) = 96$	

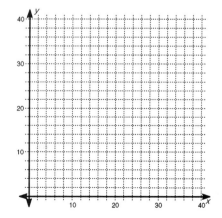
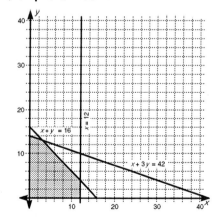

2. Graph the feasible region and determine the coordinates of its vertices. Maximize the objective function $P = 35x + 31y$ and minimize the objective function $C = 4x + 12y$ by checking its values at the vertices.

$$\begin{cases} x + 3y \geq 30 \\ -3x + 11y \leq 220 \\ 9x - 4y \leq 208 \\ 3x + 5y \geq 196 \\ x \geq 0, \ y \geq 0 \end{cases}$$

Graph and label each restriction. Labeling the edges helps determine which restrictions to use to find each vertex. Find the vertices.

$x = 0$ and $x + 3y = 30$ yields the vertex $(0, 10)$

$x = 0$ and $-3x + 11y = 220$ yields the vertex $(0, 20)$

$-3x + 11y = 220$ and $3x + 5y = 196$ yields the vertex $(22, 26)$

$3x + 5y = 196$ and $9x - 4y = 208$ yields the vertex $(32, 20)$

$9x - 4y = 208$ and $x + 3y = 30$ yields the vertex $(24, 2)$

Maximize

Vertices	Objective Function:	$P = 35x + 31y$
(0, 10)	$35(0) + 31(10) = 310$	
(0, 20)	$35(0) + 31(20) = 620$	
(22, 26)	$35(22) + 31(26) = 1576$	
(32, 20)	$35(32) + 31(20) = 1740$	Maximize
(24, 2)	$35(24) + 31(2) = 1182$	

Minimize

Vertices	Objective Function:	$C = 4x + 12y$
(0, 10)	$4(0) + 12(10) = 120$	Minimum
(0, 20)	$4(0) + 12(20) = 240$	
(22, 26)	$4(22) + 12(26) = 400$	
(32, 20)	$4(32) + 12(20) = 368$	
(24, 2)	$4(24) + 12(2) = 120$	Minimum

Since the minimum occurs at two vertices, any point along the edge joining these vertices will be a minimum.

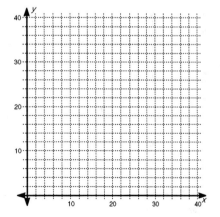
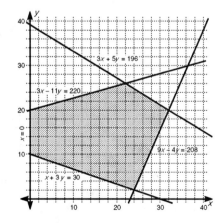

3. A farmer plans to put in a peach orchard and a cherry orchard. Each new peach tree cost $6.00 to buy and plant and requires 132 gallons of water a year to maintain. Each cherry tree cost $7.50 to buy and plant and requires 95 gallons of water a year to maintain. Suppose the farmer has $40,500 spend on the trees and water rights for 10,000,000 gallons of water a year. If he expects to make $45 per peach tree and $47 per cherry tree, how many of each should he plant?

The objective function here is profit and the farmer wants to maximize.
Let x = number of peach trees he should buy;
Let y = number of cherry trees he should buy.

Maximize $P = 45x + 47y$
Subject to:
Cost: $\quad 6x + 7.5y \leq 40{,}500$
Water: $\quad 132x + 95y \leq 660{,}000$
Nonnegative: $\quad x \geq 0, y \geq 0$

Graph and label each restriction. Find the vertices.

$x = 0$ and $6x + 7.5y = 40{,}500$ yields the vertex $(0, 5400)$

$6x + 7.5y = 40{,}500$ and $132x + 95y = 660{,}000$ yields the vertex $(2625, 3300)$

$y = 0$ and $132x + 95y = 660{,}000$ yields the vertex $(5000, 0)$

Vertices	Maximize: $P = 45x + 47y$	
(0, 0)	$45(0) + 47(0) = 0$	
(0, 5400)	$45(0) + 47(5400) = 253{,}800$	
(2625, 3300)	$45(2625) + 47(3300) = 273{,}225$	Maximum
(5000, 0)	$45(5000) + 47(0) = 225{,}000$	

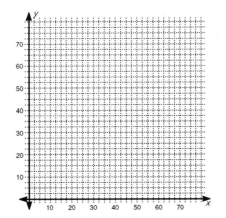
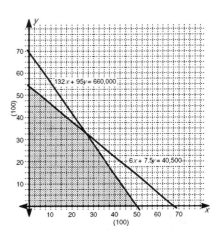

Section 7.10   Applications: Partial Fractions

Key Ideas
A. Concepts of partial fractions, four types of denominator factors.
B. The steps of partial fraction decomposition.

A. Let r be the proper rational function $r(x) = \dfrac{P(x)}{Q(x)}$, where the degree of P is less than the degree of Q. The goal is to express a proper rational function $r(x)$ as a sum of simpler functions called **partial fractions**. Every polynomial with real coefficients can be factored completely into linear factors and irreducible quadratic factors. That is, into factors of the form $ax + b$ and of the form $ax^2 + bx + c$, where a, b, and c are real numbers. There are four possible types of factors.

1. Each linear factor that occurs *exactly once* in the denominator leads to a term of the form $\dfrac{A}{ax+b}$ (where $A \ne 0$).

2. Each irreducible quadratic factor that occurs *exactly once* in the denominator leads to a term of the form $\dfrac{Ax+B}{ax^2+bx+c}$ (where A and B are not both zero).

3. When the linear factor is repeated k times in the denominator, then the corresponding partial fraction decomposition will contain the terms:
$$\dfrac{A_1}{ax+b} + \dfrac{A_2}{(ax+b)^2} + \cdots + \dfrac{A_k}{(ax+b)^k}$$

4. When the irreducible quadratic factor is repeated m times in the denominator, then the corresponding partial fraction decomposition will contain the terms:
$$\dfrac{A_1 x + B_1}{ax^2+bx+c} + \dfrac{A_2 x + B_2}{(ax^2+bx+c)^2} + \cdots + \dfrac{A_m x + B_m}{(ax^2+bx+c)^m}$$

The number k and m in types 3 and 4 above are called the **multiplicity** of the factor.

B. The technique of partial fraction decomposition can be broken down to the following steps:

1. *If the rational function is not proper,* perform long division and then apply partial fraction decomposition to the (proper) remainder.

2. *Factor* the denominator into linear and irreducible quadratic factors. Express the rational function in its partial fraction decomposition with unknown constants as discussed in part **A**.

3. *Multiply* both sides by the common denominator. *Simplify* the result by grouping like terms.

4. *Equate* corresponding coefficients. That is, the coefficient of the x^k term on either side of the equation must be equal. As a result, this leads to a system of linear equations based on these coefficient. *Solve* the resulting system of linear equations.

Find the partial fraction decomposition of the given rational function.

1. $\dfrac{-2x + 15}{x^2 - x - 12}$

Since $x^2 - x - 12 = (x - 4)(x + 3)$

$$\dfrac{-2x + 15}{x^2 - x - 12} = \dfrac{A}{x - 4} + \dfrac{B}{x + 3}$$

Multiply both sides by the common denominator.

$-2x + 15 = A(x + 3) + B(x - 4)$
$-2x + 15 = Ax + 3A + Bx - 4B$
$-2x + 15 = (A + B)x + 3A - 4B$
$-2 = A + B$
$15 = 3A - 4B$

$\begin{array}{l} -3(-2 = A + B) \\ 15 = 3A - 4B \end{array} \Rightarrow \begin{array}{l} 6 = -3A - 3B \\ 15 = 3A - 4B \\ \hline 21 = -7B \end{array}$

So $B = -3$, substituting:
$-2 = A + B = A + (-3)$
$A = 1$

$$\dfrac{-2x + 15}{x^2 - x - 12} = \dfrac{1}{x - 4} + \dfrac{-3}{x + 3}$$

2. $\dfrac{3x^3 + 2x^2 - 3x - 3}{x^4 + x^3}$

Since $x^4 + x^3 = x^3(x + 1)$

x has multiplicity 3

$$\dfrac{3x^3 + 2x^2 - 3x - 3}{x^4 + x^3} = \dfrac{A}{x} + \dfrac{B}{x^2} + \dfrac{C}{x^3} + \dfrac{D}{x + 1}$$

Multiply both sides by the common denominator.

$3x^3 + 2x^2 - 3x - 3 = Ax^2(x + 1) + Bx(x + 1) + C(x + 1) + Dx^3$

Multiply and simplify:
$= Ax^3 + Dx^3 + Ax^2 + Bx^2 + Bx + Cx + C$

set the coefficient equal and solve.

x^3 terms: $\quad 3 = A + D$
x^2 terms: $\quad 2 = A + B$
x terms: $\quad -3 = B + C$
constants: $\quad -3 = C$

Solving: $C = -3, B = 0, A = 2$, and $D = 1$

$$\dfrac{3x^3 + 2x^2 - 3x - 3}{x^4 + x^3} = \dfrac{2}{x} + \dfrac{-3}{x^3} + \dfrac{1}{x + 1}$$

3. $\dfrac{x^5}{x^4 - 5x^2 + 4}$

Since $\dfrac{x^5}{x^4 - 5x^2 + 4}$ is not proper, do long division.

$\dfrac{x^5}{x^4 - 5x^2 + 4} = x + \dfrac{5x^3 - 4x}{x^4 - 5x^2 + 4}$

Now factor the denominator.

$x^4 - 5x^2 + 4 = (x^2 - 1)(x^2 - 4)$
$ = (x - 1)(x + 1)(x - 2)(x + 2)$

$\dfrac{5x^3 - 4x}{x^4 - 5x^2 + 4} = \dfrac{A}{x - 1} + \dfrac{B}{x + 1} + \dfrac{C}{x - 2} + \dfrac{D}{x + 2}$

$5x^3 - 4x = A(x + 1)(x^2 - 4) + B(x - 1)(x^2 - 4)$
$ + C(x^2 - 1)(x + 2) + D(x^2 - 1)(x - 2)$

Instead of writing a system of 4 equations in 4 unknowns, we use substitution to find the values $A, B, C,$ and D.

If $x = -1$, then the equation becomes:

$5(-1)^3 - 4(-1) = 0A + B(-2)(-3) + 0C + 0D$

So $-5 + 4 = 6B \qquad B = -\dfrac{1}{6}$

If $x = 1$, then the equation becomes:

$5(1)^3 - 4(1) = A(2)(-3) + 0B + 0C + 0D$

So $5 - 4 = -6A \qquad A = -\dfrac{1}{6}$

If $x = 2$, then the equation becomes:

$5(2)^3 - 4(2) = 0A + 0B + C(3)(4) + 0D$

So $40 - 8 = 12C \qquad C = \dfrac{32}{12} = \dfrac{8}{3}$

If $x = -2$, then the equation becomes:

$5(-2)^3 - 4(-2) = A + 0B + 0C + D(3)(-4)$

So $-40 + 8 = 12D \qquad B = -\dfrac{32}{12} = -\dfrac{8}{3}$

$\dfrac{x^5}{x^4 - 5x^2 + 4} = x + \dfrac{5x^3 - 4x}{x^4 - 5x^2 + 4}$

$= x + \dfrac{-\frac{1}{6}}{x - 1} + \dfrac{-\frac{1}{6}}{x + 1} + \dfrac{\frac{8}{3}}{x - 2} + \dfrac{-\frac{8}{3}}{x + 2}$

$= x - \dfrac{1}{6(x - 1)} - \dfrac{1}{6(x + 1)} + \dfrac{8}{3(x - 2)} - \dfrac{8}{3(x + 2)}$

Chapter 8

Counting and Probability

Section 8.1   Counting Principles
Section 8.2   Permutations
Section 8.3   Combinations
Section 8.4   Problem Solving With Permutations and Combinations
Section 8.5   Probability
Section 8.6   The Union of Events
Section 8.7   The Intersection of Events
Section 8.8   Expected Value

Section 8.1   Counting Principles

Key Ideas
A. Tree Diagrams.
B. Fundamental counting principle.
C. Factorial notation.
D. Counting with cards and dice.

A. A **tree diagram** is a systematic way of finding and listing all possible results of a counting or probability problem. At each vertex, all possible choices are drawn out a edges and taken. To create a list of possible outcomes to a counting problem, start at an end vertex and trace your way back to the initial vertex.

1. A family is planning on having three children, list all possible orders of birth.

    Here we use a tree diagram to help use count. Let B represent a boy and G represent a girl..

    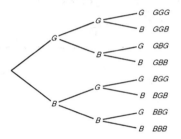

    So the 8 possible child births are:
    GGG, GGB, GBG, GBB, BBG, BBB, BGG, BGB

2. A pet food manufacturer test two types of dog food to see which food is preferred by a dog. A dog is observed <u>at most</u> five times or until it chooses the same food three times. How many different outcomes are possible? Draw a tree diagram to count this question.

    Although this count can be made with other methods to come, using a tree diagram is the easiest. Let A and B represent the two types of dog food. Draw the tree.

    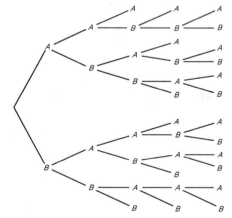

    Solution: There are 20 possible outcomes.

B. The **fundamental counting principle** states that if two events occurs in order and if the first event can occur in n ways and the second in m ways (after the first event has occurred), then the two events can occur in order in $n \times m$ ways. When the events $E_1, E_2, \ldots, E_k$ occur in order and if E_1 can occur in n_1 ways, E_2 in n_2 ways, and so on, then the events can occur in order in $n_1 \times n_2 \times \cdots \times n_k$ ways.

3. For security purposes, a company classifies each employee according to five hair colors, four eye colors, six weight categories, five height categories, and two sex categories. How many classifications are possible?

   Using the fundamental counting principle there are: $\underline{5} \cdot \underline{4} \cdot \underline{6} \cdot \underline{5} \cdot \underline{2} = 1200$ classifications.

4. The Uniform Pricing Code, the bars on the side of packages that stores use for scanning the item number and then printing the price, consists of a 10 digit number. How many different bars are possible?

   Using the fundamental counting principle there are ten choices for each of the ten digits. As a result there are:
   $\underline{10} \cdot \underline{10} \cdot \underline{10} \cdot \underline{10} \cdot \underline{10} \cdot \underline{10} \cdot \underline{10} \cdot \underline{10} \cdot \underline{10} \cdot \underline{10} = 10^{10}$
   Uniform Pricing Codes

5. A car manufacturer makes its lowest priced car in 5 exterior colors, 4 interior colors, a choice of manual or automatic transmission, and 3 levels of options. How many basic models are available?

   Using the fundamental principle of counting, there are:
   $\underline{5} \cdot \underline{4} \cdot \underline{2} \cdot \underline{3} = 120$ basic models available.

C. The product of the first n natural numbers is denoted by **$n!$** and is called **n factorial**.
$n! = 1 \cdot 2 \cdot 3 \cdot 4 \cdots (n-1) \cdot n$, where $0! = 1$
n factorial can also be expressed as $n! = n \cdot (n-1)!$ for $n \geq 1$. For example $6! = 6 \cdot (5!)$

6. A woman sits down to make four sales calls, in how many ways can she make her calls?

   She has 4 choices for the first call, 3 choices for the second call, 2 choices for the third call and 1 choices for the last call.

   So there are $4 \cdot 3 \cdot 2 \cdot 1 = 4!$ ways for her to make her calls.

7. Simplify each expression.

   a) $\dfrac{8!}{3!}$

   $\dfrac{8!}{3!} = \dfrac{8 \cdot 7 \cdot 6 \cdot 5 \cdot 4 \cdot 3 \cdot 2 \cdot 1}{3 \cdot 2 \cdot 1} = 8 \cdot 7 \cdot 6 \cdot 5 \cdot 4 = 6720$

   b) $\dfrac{15!}{5!10!}$

   $\dfrac{15!}{5!10!}$

   $= \dfrac{15 \cdot 14 \cdot 13 \cdot 12 \cdot 11 \cdot 10 \cdot 9 \cdot 8 \cdot 7 \cdot 6 \cdot 5 \cdot 4 \cdot 3 \cdot 2 \cdot 1}{5 \cdot 4 \cdot 3 \cdot 2 \cdot 1 \cdot 10 \cdot 9 \cdot 8 \cdot 7 \cdot 6 \cdot 5 \cdot 4 \cdot 3 \cdot 2 \cdot 1}$

   $= \dfrac{15 \cdot 14 \cdot 13 \cdot 12 \cdot 11}{5 \cdot 4 \cdot 3 \cdot 2 \cdot 1}$

   Using the definition $n! = n \cdot (n-1)!$

   $\dfrac{15!}{5!10!} = \dfrac{15 \cdot 14 \cdot 13 \cdot 12 \cdot 11 \cdot 10!}{5 \cdot 4 \cdot 3 \cdot 2 \cdot 1 \cdot 10!}$

   $= \dfrac{15 \cdot 14 \cdot 13 \cdot 12 \cdot 11}{5 \cdot 4 \cdot 3 \cdot 2 \cdot 1} = 3003$

   c) $\dfrac{(n+1)!}{(n-1)!}$

   Using the definition $n! = n \cdot (n-1)!$

   $\dfrac{(n+1)!}{(n-1)!} = \dfrac{(n+1)(n)(n-1)!}{(n-1)!}$

   $= (n+1)n = n^2 + n$

D. Many counting questions involve a standard deck of cards. A standard deck of cards consist of 52 cards broken up into 4 **suits** of 13 cards each. The suits **diamonds** and **hearts** are red, while the suits **clubs** and **spades** are black. The cards *3 of diamonds, 3 of hearts, 3 of clubs* and *3 of spades* form a **kind**. There are 13 kinds in a deck, Ace, 2, 3, ... 10, Jack, Queen, King.

Another common counting questions involve a **die** (or **dice**–plural) is a six-sided cube with the numbers 1 through 6 painted on faces. When a pair of dice is rolled there are 36 possible outcomes that can occur. Also, a common "experiment" is to look at the sum of the faces. The table on the right shows the 36 possible outcomes along with the corresponding sum of the faces.

(G,R) sum		1	2	Red 3	4	5	6
	1	(1,1) 2	(1,2) 3	(1,3) 4	(1,4) 5	(1,5) 6	(1,6) 7
	2	(2,1) 3	(2,2) 4	(2,3) 5	(2,4) 6	(2,5) 7	(2,6) 8
Green	3	(3,1) 4	(3,2) 5	(3,3) 6	(3,4) 7	(3,5) 8	(3,6) 9
	4	(4,1) 5	(4,2) 6	(4,3) 7	(4,4) 8	(4,5) 9	(4,6) 10
	5	(5,1) 6	(5,2) 7	(5,3) 8	(5,4) 9	(5,5) 10	(5,6) 11
	6	(6,1) 7	(6,2) 8	(6,3) 9	(6,4) 10	(6,5) 11	(6,6) 12

Section 8.2 Permutations

Key Ideas
A. Permutations.
B. Distinguishable Permutations.

A. A **permutation** of a set of distinct objects is an *ordering* or *arrangment* of these objects. The number of ways n objects can be arranged is $n!$ When only r of the n objects of the set are arranged then **the number of permutations of n objects taken r at a time** is denoted by $P(n, r)$ and is given by the formula

$$P(n, r) = n \cdot (n-1) \cdot (n-2) \cdots (n-r+1) = \frac{n!}{(n-r)!}.$$

Permutations are used when (1) order is important and (2) repeats are not allowed.

1. A club of 20 people gather for a group photo standing in a row. In how many ways can the photo be taken?

   Here we are after all 20 person arrangement of the group. So there are $P(20, 20) = 20!$ ways to arrange the group.

   $20! \approx 2.432902008 \times 10^{18}$

2. Five couples line up outside a movie theater. In how many ways can they line up?

   First treat each couple as one item and arrange the couples; there are $P(5, 5) = 5!$ ways to do this. Then arrange the two people in *each* couple; there $P(2, 2) = 2!$ ways to arrange each couple.

   So there are $5! \, (2!)^5$ ways to arrange the couples in line at the movie theater.

   $5! \, (2!)^5 = 5! \, 2^5 = 3840$

3. A volleyball team has nine members and six members are on the court at any one time. In how many ways can the six be selected and arrange on the court?

   $P(9, 6) = \dfrac{9!}{3!} = \dfrac{9 \cdot 8 \cdot 7 \cdot 6 \cdot 5 \cdot 4 \cdot 3!}{3!} = 60480$

B. When considering a set of objects some of which are of the same kind, then two permutations are **distinguishable** if one cannot be obtained from the other by interchanging the positions of elements of the same kind. If a set of n objects consists of k different kinds of objects with n_1 objects of the first kind, n_2 objects of the second kind, n_3 objects of the third kind, and so on, where $n_1 + n_2 + n_3 + \cdots + n_k = n$, then the number of distinguishable permutations of these objects is $\frac{n!}{n_1! n_2! n_3! \cdots n_k!}$. The numbers $n_1, n_2, n_3, \ldots, n_k$ are called a **partition** of n when $n_1 + n_2 + n_3 + \cdots + n_k = n$.

4. An instructor makes three versions of an exam for her chemistry class with 30 students. If she makes 10 copies of each version, in how many ways can she distribute the exams in her class?

   Since all 30 exams are distributed to the class, this is a distinguishable permutation. The number of ways is then:
   $$\frac{30!}{10!\ 10!\ 10!} = 5.550996791 \times 10^{12}$$

5. How many ways can the letters of the word "MOTORSPORTS" be arranged?

   Since MOTORSPORTS has 11 letters, 1 m's, 3 o's, 2 t's, 2 r's, 2 s's, and 1 p's, there are
   $$\frac{11!}{1!\ 3!\ 2!\ 2!\ 2!\ 1!} = 831{,}600 \text{ ways to arrange these letters.}$$

6. An ice chest contains six cokes, four orange sodas, eight fruit flavored waters. In how many ways can these be distributed to 18 people at a picnic?

   Since we distribute all 18 items in the ice chest this is a distinguishable permutation. So there are
   $$\frac{18!}{6!\ 4!\ 8!} = 9189180 \text{ was to distribute the soft drinks.}$$

Section 8.3 Combinations

Key Ideas
A. Combinations.
B. Distinguishable Combinations.

A. A **combination** of r elements of a set is any subset of r elements from the set (*without regard to order*). If the set has n elements then the **number of combinations of n elements taken r at a time** is denoted by $C(n, r)$ and is given by the formula
$$C(n, r) = \frac{n!}{r!\,(n-r)!}.$$
Combinations are used when (1) order is *not* important and (2) there are no repeats.

1. A drawer contains 28 pairs of socks, in how many ways can a person select ten to pack for a vacation?

    Order is not important, no repeats. So there are
    $$C(28,10) = \frac{28!}{10!\,(28-10)!} = \frac{28!}{10!\,18!} = 13123110$$
    ways to selected the pairs of socks.

2. Twelve cars are stopped at a fruit inspection checkpoint operated by the state of California. In how many ways can the inspectors choose three of the twelve cars to inspect?

    Order is not important because they are after a *group* of cars to inspect.
    No repeats since you will not inspect a car twice.
    So there are $C(12, 3) = \dfrac{12!}{3!\,(12-3)!} = \dfrac{12!}{3!\,9!} = 220$
    ways to selected the cars to inspect.

3. A group of 12 friends are at a party.
    a) In how many ways can 3 people be selected to go out and get more soft drinks?

    Since a group is selected, order is not important, and obvious that there are no repeats. As a result
    there are $C(12, 3) = \dfrac{12!}{3!\,(12-3)!} = \dfrac{12!}{3!\,9!} = 220$ ways
    three people can be selected.

    b) If there are 6 females and 6 males at the party, in how many ways can 2 males and one female be selected to get the soft drinks?

    This is still a combination. Since the males can only be selected from the six males and the female can only be selected from the six females, there are
    $$C(6, 2) \cdot C(6,1) = \frac{6!}{2!\,4!} \cdot \frac{6!}{1!\,5!} = 15 \cdot 6 = 90 \text{ ways to}$$
    select the three people.

4. Pauline's Pizza Palace has 14 different kinds of toppings. Pauline's Special Combination is a large pizza with any three different toppings. How many Pauline's Special Combinations are possible?

Since these toppings must be different and order is not important, this is a combination problem. The number of Pauline's Special Combinations is
$$C(14, 3) = \frac{14!}{3!\,11!} = 364.$$

B. When a collection of objects is made up of groups of the like kind then two combinations are **indistinguishable** if they contain the same number of objects of each kind. Otherwise, they are **distinguishable**. If a set of n objects consists of k different kinds of objects with n_1 objects of the first kind, n_2 objects of the second kind, n_3 objects of the third kind, and so on, where $n_1 + n_2 + n_3 + \cdots + n_k = n$, then the number of distinguishable combinations of these objects is $(n_1 + 1)(n_2 + 1)(n_3 + 1) \cdots (n_k + 1)$. A special case happens when a consider a general set has n elements, using $n_i = 1$, for $1 \leq i \leq n$, we obtain the result that the set has 2^n subsets.

5. A shopping bag containing 12 oranges, 10 apples, and 8 peaches tips over and some of the contents slip out. How many different groups of fruit can be left in the bag?

What is left in the bag is a subset of what was in the bag to begin with. Since the number of subsets is $(12+1)(10+1)(8+1) = 13 \cdot 11 \cdot 9 = 1287$ and we are told that *some of the contents slip out* we have to remove the case where nothing slips out. So the number of different groups that could be left in the bag is $1287 - 1 = 1286$.

6. A hamburger chain advertises that they can make a hamburger in 256 ways. How many different toppings do they need to meet this claim?

Since changing a topping will change the type of hamburger, a subset of toppings will represent a different hamburger. So we need to solve the equation $2^n = 256$.
Thus $n = \log_2 256 = 8$
The hamburger chain needs 8 toppings.

Section 8.4   Problem Solving With Permutations and Combinations

Key Ideas
A. Difference between permutations and combinations.

A. Both permutations and combinations use a set of n distinct objects where r objects are selected at random *without repeats* (or replacements). The key difference between permutations and combinations is *order*. When order *is important*, **permutations** are used and when order *is not* important, **combinations** are used. Other key words that help to distinguish combinations are: *group, sets, subsets, etc.*. Key words that distinguish permutations are: *arrange, order, arrangements, etc.*

1. Five science students study for a math quiz.
   a) In how many ways can they sit around a round table in the library?

   First fix one student as a reference point and arrange the others around this student. There are $4! = 24$ ways they can sit.

   b) Suppose two students *must* sit next to each other. Now, in how many ways can they sit around a round table in the library?

   First treat the two students that *must* sit next to each other as a single unit. As in the question above, first fix one of the 4 items to be arranged. So there are 3! ways to sit them around the table. Now, there are 2! ways to arrange the two students that *must* sit next to each other. As a result there are $2! \cdot 3! = 12$ ways for them to sit around the table.

2. Fifteen people are selected for a study of diet pills. Eight will be given the new diet pill and seven will be given a placebo. In how many ways can the eight be selected?

   This is a combination problem because (1) there are no repeats, and (2) you are interested in a group only.

   So there are $C(15, 8) = \dfrac{15!}{8!\, 7!} = 6435$ ways to select the test group.

3. A college theater group plans on doing 5 different shows during the year- *Cats, The Tempest, Annie, Streetcar Named Desire* and *Guys and Dolls*. If *Cats* can only be scheduled last, and *Annie* must be either first or second, how many different schedules are possible?

There are five slots to fill; *Cats* must be placed in the last slot, while *Annie* can be placed in either the first or second slot. So there is 2 ways to place *Annie* and 1 way to place *Cats* (in the last slot) and 3 remaining slots to fill. Since there are 3! ways to fill these remaining slots there are a total of $2 \cdot 3!$ to schedule the shows.

4. A newly formed Ad Hoc Committee on Student Retention plans to study ten new students during the Fall semester to identify ways to improve student retention. Suppose next fall's incoming class contains 112 new female and new 108 male students.
   a) In how many different ways can 10 new students be chosen at random (with no regard to sex) for this study?

   Since there are 220 new students expected, we pick the 10 students from this group.
   So there are $C(220, 10) = \dfrac{220!}{10!\,210!}$ ways to pick the study group.

   b) How many of these possible studies in part **a** have 5 females and 5 males in them?

   The five female students can only be picked from the 112 female students and the five male students can only be picked from the 108 students.
   So there $C(112, 5) \cdot C(108, 5) = \dfrac{112!}{5!\,107!} \cdot \dfrac{108!}{5!\,103!}$ ways to form the study group.

B. Hands of cards are used in many counting questions. The following problems deal with card hands that have special names. For a explanation of a standard deck of cards see page 207 of this study guide.

5. A *full house* is any three cards from one kind and any two cards from a second kind. How many different full hours are possible?

    We solve this problem by using the fundamental counting principle. In words, to get a full house you must get three cards of one kind and two cards from a second kind.

    $$\binom{kind}{1}\binom{3 \text{ of } 4}{cards}\binom{kind}{2}\binom{2 \text{ of } 4}{cards}$$

    $= C(13, 1) \cdot C(4, 3) \cdot C(12, 1) \cdot C(4, 2)$

    $= \dfrac{13!}{1! \cdot 12!} \dfrac{4!}{3! \cdot 1!} \dfrac{12!}{1! \cdot 11!} \dfrac{4!}{2! \cdot 2!} = 13 \cdot 4 \cdot 12 \cdot 6$

    $= 3744$

6. A *flush* is five cards form the same suit. How many different flushes are possible?

    Here we first pick the suit and then 5 cards from that suit.
    $C(4, 1)$ ways to pick the suit
    $C(13, 5)$ ways to pick the 5 cards from the suit.
    So there are $C(4, 1) \cdot C(13, 5) = 5148$ flushes.

7. A *straight* is any five cards in a row, from Ace-2-3-4-5 to 10–Jack-Queen-Kings-Ace. How many different straights are possible?

    This is a harder question. We first count the number of different straights that are possible and then chose suits for the cards.
    The different straights can be determine by the card they start with (the lowest card in the straight). Remember an Ace can be both low and high. So straights can start Ace to those who start with a 10. 10 different straights.

    Now there are $C(4, 1) = 4$ ways to pick a suit for each card, and 5 cards in the hand.
    So there are $10 \cdot 4^5 = 10240$ straights

    Of these, 40 are straight flushes (pick a straight and then a suit), 1024 are royal straights (10–Jack-Queen-Kings-Ace), and 4 royal straight flushes.

Section 8.5   Probability

Key Ideas
A. Sample spaces and probability.
B. Complements.

A. An **experiment** is process that gives definite outcomes. The set of all possible outcomes of experiment is called the **sample space**. An **event** is any subset of the sample space. When each individual outcome has an equal chance of occurring, then the probability that event E occurs, written as $P(E)$, is

$$P(E) = \frac{n(E)}{n(S)} = \frac{\text{number of elements in event } E}{\text{number of elements in sample space } S}$$

As a result, $0 \leq P(E) \leq 1$. The closer $P(E)$ is to one the more likely the event is to happen, while the closer to zero the less likely. If $P(E) = 1$, then E is called the **certain event**, that means E is *the only event that can happen*. When $P(E) = 0$ the event is called an **impossible event**.

1. Let S be the experiment where a pair of dice are rolled and the sum of the faces recorded.

   a) Find the probability that the sum of the numbers showing is 3.

   Using the table on page 207 of this study guide, we see that the sum is 3 in only two of the 36 squares.
   $$P(\text{sum is 3}) = \frac{2}{36} = \frac{1}{18}$$

   b) Find the probability that the sum of the faces is odd.

   Again looking the table on page 207, there are 18 squares where the sum is odd, so the probability is
   $$P(\text{sum is odd}) = \frac{18}{36} = 0.5$$

   c) Find the probability that the sum of the faces is greater than 9.

   Sum is greater than 9 is the same as the sum is 10 or the sum is 11 or the sum is 12. looking at the table, there are 3 squares where the sum is 10, 2 squares where the sum is 11, and 1 square where the sum is 12 for a total of 6 squares.
   $$P(\text{sum is greater than 9}) = \frac{6}{36} \approx 0.167$$

2. A box of 100 ping pong balls contains 50 yellow and 50 white balls. If 5 balls are selected at random, what is the probability that
  a) all 5 balls are yellow?

Since there are 100 balls, the sample space, S, consists of all ways to select 5 balls without regard to color. So $n(S) = C(100, 5) = 75287520$.
Since the yellow balls can only be selected from the 50 yellow balls and there are $C(50, 5) = 2118760$

$$P(5 \text{ yellow balls}) = \frac{C(50, 5)}{C(100, 5)} = \frac{2118760}{75287520} = 0.02814$$

  b) 3 balls are yellow and 2 balls are white?

Pick the 3 yellow balls from the 50 yellow balls and pick the 2 white balls from the 50 white balls.
$C(50,3) \cdot C(50,2) = 19600 \cdot 1225 = 24010000$
So the probability is

$$P(3 \text{ yellow}, 2 \text{ white}) = \frac{C(50,3) \cdot C(50,2)}{C(100, 5)} = \frac{24010000}{75287520}$$
$$= 0.3189$$

B. The **complement** of an event E is the set of outcomes of the *sample space* S that are not in E, and is denoted as E'. The probability that E *does not occur*, $P(E') = 1 - P(E)$. Sometimes it is easier to find $P(E')$ than to count find $P(E)$.

3. A box of 100 calculators contains 3 defective calculators. Suppose five calculators are selected, what is the probability that at least one calculator is defective?

The complement of *at least one defective* is *no defective* which is easier to count.
There are 97 non-defectives chose 5, $C(97,5)$

$$P(\text{non-defective}) = \frac{C(97,5)}{C(100,5)} = 0.8738$$

P(at least 1 defective) = 1 − P(non-defective)
  = 1 − 0.8738 = 0.1262

4. Five friends compare birth months. What is the probability that at least two will be born in the same month?

Here we assume that each month is equally likely to occur. Since we seek *at least two* this means that two could have the same birth month, three could have the same birth month, two pairs could have the same birth month, etc. The complement, *no two have the same birth month*, is much easier to count. P(different birth months)

$$= \frac{\text{number of ways to assign 5 different birth months}}{\text{number of ways to assign birth months}}$$

$$= \frac{12 \cdot 11 \cdot 10 \cdot 9 \cdot 8}{12 \cdot 12 \cdot 12 \cdot 12 \cdot 12} = \frac{55}{144}$$

So the probability that at least two will have the same birth month is

$$1 - \frac{55}{144} = \frac{89}{144} \approx 0.618$$

Section 8.6   The Union of Events

Key Ideas
A. Mutually exclusive events and unions of events.
B. Probability of the union of two sets.

A. When two events have no outcomes in common they are said to be **mutually exclusive**. The probability of the union of two *mutually exclusive* sets E and F in a sample space S is given by $P(E \cup F) = P(E) + P(F)$

1. Box contains 6 red building block and 4 blue building blocks. An infant selects 5 blocks from the box. Are the events E, the infant selects 3 red and 3 blue building blocks, and F, the infant selects all red building blocks, mutually exclusive?

   Since blue building blocks are not red, an outcome cannot satisfy both events, so $E \cap F = \emptyset$ and the events are mutually exclusive.

2. A five card hand is drawn. Consider the events: E - all 5 cards are red; F - full house; G - two pairs; H - all 5 cards are faces cards. Which of these events are mutually exclusive?

   E and F : These events are mutually exclusive because a full house means that there are 3 cards of one kind, at least 1 of these 3 cards cannot be red. So a full house has to have at least one black card.

   E and G : These events are not mutually exclusive. There are many hands where you have 2 pairs and a fifth card that are all red.

   E and H : These events are not mutually exclusive. There are three face cards per suit, so there are six red face cards and all we need is five.

   F and G : By definition two pairs means that the fifth card is from a different kind that the pairs. As a result these events are mutually exclusive.

   F and H : These events are not mutually exclusive. There are many full houses which involve only face cards.

   G and H : These events are not mutually exclusive. Since there are three kinds of face cards, each pair and the fifth card can come from a different kind.

B. When two events are *not* mutually exclusive, the probability of E or F is
$$P(E \cup F) = P(E) + P(F) - P(E \cap F)$$
Remember, when the word **or** is used, you are looking for the union of the events.

3. A five card hand is drawn at random. Find the the probability that all red cards are drawn or a full house is drawn.

    The sample space is all 5 card hands, $C(52, 5)$.
    Let E = all 5 cards are red.
    Let F = full house.
    Using the events from Problem 2 above, we saw that these events are mutually exclusive.
    Since there are 26 red cards from which 5 red cards is drawn $P(E) = \dfrac{C(26, 5)}{C(52, 5)}$.

    $P(F) = \dfrac{3744}{C(52, 5)}$ (see Problem 5, page 214)

    $P(E \cup F) = P(E) + P(F) = \dfrac{C(26, 5)}{C(52, 5)} + \dfrac{3744}{C(52, 5)}$

4. A five card is drawn at random. Find the probability that a full house is drawn or all five cards are face cards.

    The sample space is all 5 card hands, $C(52, 5)$.
    Let F = full house.
    Let H = all five cards are face cards.
    Using the events from Problem 2 above, we saw that these events are not mutually exclusive.

    $P(F) = \dfrac{3744}{C(52, 5)}$

    Each suit has 3 face cards and there are 4 suits, so there are 12 face cards, chose 5; so $n(H) = C(12, 5)$
    and $P(H) = \dfrac{C(12, 5)}{C(52, 5)} = \dfrac{792}{C(52, 5)}$

    $F \cap H$ is the event that the full house is made from face cards. Since there are 3 kinds of face cards, there are $C(3, 1)$ to pick the kind and $C(4, 3)$ ways to pick 3 of the 4 cards from the kind, $C(2, 1)$ ways to pick the second kind and $C(4, 2)$ ways to pick 2 of the 4 cards from that kind.

    $n(F \cap H) = C(3, 1) \cdot C(4, 3) \cdot C(2, 1) \cdot C(4, 2)$
    $= 3 \cdot 4 \cdot 2 \cdot 6 = 144$

    $P(F \cap H) = \dfrac{144}{C(52, 5)}$

    $P(F \cup H) = P(F) + P(H) - P(F \cap H)$
    $= \dfrac{3744}{C(52, 5)} + \dfrac{792}{C(52, 5)} - \dfrac{144}{C(52, 5)}$
    $= \dfrac{4392}{C(52, 5)}$

Section 8.7 The Intersection of Events

Key Ideas
A. Conditional probability.
B. Independent events.

A. Often when considering probabilities there is additional information available that affects the calculation of the probability of an event. If E and F are events in a sample space S, then the **conditional probability** that F occurs given E has occurred is given by the formula $P(F/E) = \dfrac{P(E \cap F)}{P(E)}$. In words, $P(F/E)$ says that when you *know* that event E has occurred, then event E becomes the new sample space. When each event in the sample space is equally likely to occur, then conditional probability is given by the formula $P(F/E) = \dfrac{n(E \cap F)}{n(E)}$. The equation $P(F/E) = \dfrac{P(E \cap F)}{P(E)}$ can be manipulated algebraically into the form $P(E \cap F) = P(E) \cdot P(F/E)$. In words, this says that you can get into the intersection $E \cap F$ by first getting into E and then getting into F given you are already in E.

1. A red die and a green die are rolled.
    a) What is the probability that the sum of the faces is 9 given the green die is a 4?

    Let F be the event that sum of the faces is 9.
    Let E = event that the green die shows a 4.
    Since each event in this sample space is equally likely to occur, we use the formula
    $$P(F/E) = \frac{n(E \cap F)}{n(E)} = \frac{1}{6}$$

    b) What is the probability that the sum of the faces is 9 given the green die is a 1?

    Let F be the event that sum of the faces is 9.
    Let E = event that the green die shows a 1.
    Since each event in this sample space is equally likely to occur, we use the formula
    $$P(F/E) = \frac{n(E \cap F)}{n(E)} = \frac{0}{6} = 0$$

2. An urn contains 4 blue balls and 3 green balls. Two balls are drawn in order from the urn.
    a) What is the probability that the second ball is green given the first ball was green?

    Let G_i be the event that the ith ball selected is green. $P(G_2/G_1) = \dfrac{\frac{P(4,\ 2)}{P(7,\ 2)}}{\frac{P(4,\ 1)}{P(7,1)}} = \dfrac{\frac{2}{7}}{\frac{4}{7}} = \dfrac{1}{2}$

    Since the first ball was green, on the second pick the urn has 3 green balls and 3 blue balls, so
    $P(G_2/G_1) = \dfrac{3}{6} = \dfrac{1}{2}$

219

b) What is the probability that the second ball is green given the first ball was green?

Let G_2 be the event that the 2nd ball selected is green and let B_1 be the event that the first ball is blue. Since the first ball was red, on the second pick the urn has 4 green balls and 2 blue balls, so
$$P(G_2/B_1) = \frac{4}{6} = \frac{2}{3}$$

Using $P(G_2/B_1) = \dfrac{\dfrac{P(3,1)\ P(4,1)}{P(7,2)}}{\dfrac{P(3,1)}{P(7,1)}} = \dfrac{\dfrac{2}{7}}{\dfrac{3}{7}} = \dfrac{2}{3}$

3. Four roommates draw straws to see who washes the dishes. One short straw is placed with 3 long straws.

a) What is the probability that the first straw draw is short?

Since there is 1 short straw in 4 straws,
$$P(\text{short}) = \frac{1}{4}$$

b) What is the probability that the second straw draw is short given the first straw was long?

If the first straw is long, then there is 1 short straw in 3 straws, so $P(\text{short/long}) = \dfrac{1}{3}$

c) What is the probability that the third straw draw is short given the first two straws where long?

If the first two straws is long, then there is 1 short straw in 2 straws, so $P(\text{short/long}) = \dfrac{1}{2}$

B. Two events are **independent** when the occurrence of one event does not affect the probability of another event occurring. Event E and F are independent if $P(E/F) = P(E)$ and $P(F/E) = P(F)$. In words this says, that E occurring does not influence F occurring. When events E and F are independent, then $P(E \cap F) = P(E) \cdot P(F)$. Many students get *independent events* and *mutually exclusive events* confused. The difference is that when two events are mutually exclusive, they have nothing in common but when the events are independent it says that the events do not influence each other. If two events are independent they are *not* mutually exclusive.

4. A pair of dice are rolled 5 times. What is the probability of rolling five 7's in a row?

Since rolling a 7 on any one row does not influence the next roll, this probability is given by

$$\frac{1}{6} \cdot \frac{1}{6} \cdot \frac{1}{6} \cdot \frac{1}{6} \cdot \frac{1}{6} = \frac{1}{6^5}$$

5. A driver on a certain stretch of interstate has a 0.001 probability of getting a radar ticket. Suppose a driver drives this stretch of road 6 times, what is the probability that the driver gets at least one radar ticket?

Since the complement is easier to find, we find it first.
The complement of getting at least one ticket is getting NO tickets.
P(no ticket) = 1 − 0.001 = 0.999

So $(0.999)^6$ = 0.994 is the probability of no tickets in 6 trips and the P(getting a ticket) = 1− 0.994 =0.006

Section 8.8 Expected Value

Key Ideas
A. What Is Expected Value?

A. The **expected value** of an experiment is a weighted average of *payoffs* based on the outcomes of the experiment. The weight applied to each *payoff* is the probability of the event (that corresponds to the payoff) occurring. Expected value E is
$$E = x_1 p_1 + x_2 p_2 + \ldots + x_n p_n$$
Here x_1 is the payoff and p_i is the probability.

1. A game consists of rolling a die. You receive the face value if an odd number is rolled. If the game costs $1.50 to play, what is the expected value to you?

    The payoffs are:

Payoff	EventDescription	Probability
−$1.50	an even number is rolled	$\frac{1}{2}$
−0.50	a one is rolled (−1.50 + 1)	$\frac{1}{6}$
1.50	a 3 is rolled (−1.50 + 3)	$\frac{1}{6}$
3.50	a 5 is rolled (−1.50 + 5)	$\frac{1}{6}$

    $$E = -\$1.50 \cdot \frac{1}{2} - 0.50 \cdot \frac{1}{6} + 1.50 \cdot \frac{1}{6} + 3.50 \cdot \frac{1}{6} = 0$$

2. A sales clerk at a car stereo store works for commission only. If 20% of the people will purchase an item with $3 commission and 10% will purchase a item with $5 commission and 5% will purchase an item with $8 commission, what is the expected value per customer who enters the store?

    The payoffs (commissions) are:

Payoff	Probability
$3	0.20
$5	0.10
$8	0.05
$0	the rest

    $E = 3(0.20) + 5(0.10) + 8(0.05) + 0 = 0.6 + 0.5 + 0.4$
    $= 1.5$

3. A bowl contains eight chips which cannot be distinguished by touch alone. Five chips are marked $1 each and the remaining three chips are marked $4 each. A player is blindfolded and draws two chips at random without replacement from the bowl. The player is then paid the sum of the two chips. Find the expected payoff of this game.

Here the payoff is $2 (2 $1 chips are selected), $5 ($1 chip and a $4 chip) and $8 (2 $4 chips are selected)

The probability of $2 is $\dfrac{C(5, 2)}{C(8, 2)} = \dfrac{5}{14}$

The probability of $5 is $\dfrac{C(5, 1) \cdot C(3, 1)}{C(8, 2)} = \dfrac{15}{28}$

The probability of $8 is $\dfrac{C(3, 2)}{C(8, 2)} = \dfrac{3}{28}$

$$E = 2 \cdot \dfrac{5}{14} + 5 \cdot \dfrac{15}{28} + 8 \cdot \dfrac{3}{28}$$
$$= \dfrac{119}{28} = \$4.25$$

4. A box of 10 light bulbs contains 3 defectives. If a random sample of 2 bulbs is drawn from the box, what is the expected number of defective bulbs?

Here the payoff is defective bulbs.

Payoff	Probability
0	$\dfrac{C(7,2)}{C(10,2)} = \dfrac{7}{15}$
1	$\dfrac{C(7,1)\ C(3,1)}{C(10,2)} = \dfrac{7}{15}$
2	$\dfrac{C(3,2)}{C(10,2)} = \dfrac{1}{15}$

$$E = 0\,\dfrac{7}{15} + 1\,\dfrac{7}{15} + 2\,\dfrac{1}{15} = \dfrac{9}{15} = \dfrac{3}{5}$$

Chapter 9

Sequences and Series

Section 9.1   Sequences
Section 9.2   Arithmetic and Geometric Sequences
Section 9.3   Series
Section 9.4   Arithmetic and Geometric Series
Section 9.5   Annuities and Installment Buying
Section 9.6   Infinite Geometric Series
Section 9.7   Mathematical Induction
Section 9.8   The Binomial Theorem

Section 9.1 Sequences

Key Ideas
A. General sequence.
B. Special sequences–Fibonacci sequence.
C. Finding patterns.

A. A **sequence** is a set on numbers written in a specific order:
$$a_1, a_2, a_3, a_4, \ldots, a_n, \ldots$$
The number a_1 is called the *first term*, a_2 is the *second term*, and in general a_n is the *nth term*. A **sequence** can also be defined as a function f whose domain is the set positive numbers. In this form the values $f(1), f(2), f(3), \ldots$ are called the **terms** and $F(n) = a_n$ is the *nth term*. Some sequence have formulas that give the *nth* term directly, these sequence are called **explicit**. Other sequence have formulas that give you the *nth* term based on some previous terms, when this is the case the sequence is called **recursive**.

1. Find the first five terms, the 10th term, and the 100th term of the sequence defined by the following formulas.

a) $a_n = \dfrac{n-1}{2n+1}$

$a_n = \dfrac{n-1}{2n+1}$

$n=1 \quad a_1 = \dfrac{1-1}{2(1)+1} = 0 \quad n=2 \quad a_2 = \dfrac{2-1}{2(2)+1} = \dfrac{1}{5}$

$n=3 \quad a_3 = \dfrac{3-1}{2(3)+1} = \dfrac{2}{7} \quad n=4 \quad a_4 = \dfrac{4-1}{2(4)+1} = \dfrac{3}{9}$

$n=5 \quad a_5 = \dfrac{5-1}{2(5)+1} = \dfrac{4}{11}$

$n=10 \quad a_{10} = \dfrac{10-1}{2(10)+1} = \dfrac{9}{21}$

$n=100 \quad a_{100} = \dfrac{100-1}{2(100)+1} = \dfrac{99}{201}$

b) $a_n = 2^n + (-1)^n$

$a_n = 2^n + (-1)^n$

$n=1 \quad a_1 = 2^{(1)} + (-1)^{(1)} = 2 - 1 = 1$

$n=2 \quad a_2 = 2^{(2)} + (-1)^{(2)} = 4 + 1 = 5$

$n=3 \quad a_3 = 2^{(3)} + (-1)^{(3)} = 8 - 1 = 7$

$n=4 \quad a_4 = 2^{(4)} + (-1)^{(4)} = 16 + 1 = 17$

$n=5 \quad a_5 = 2^{(5)} + (-1)^{(5)} = 32 - 1 = 31$

$n=10 \quad a_{10} = 2^{(10)} + (-1)^{(10)} = 1024 + 1 = 1025$

$n=100 \quad a_{100} = 2^{(100)} + (-1)^{(100)} = 2^{100} + 1$

c) $a_n = \dfrac{2n - 3^n}{4^n}$

$a_n = \dfrac{2n - 3^n}{4^n}$

$n = 1 \quad a_1 = \dfrac{2(1) - 3^1}{4^1} = \dfrac{2 - 3}{4} = -\dfrac{1}{4}$

$n = 2 \quad a_2 = \dfrac{2(2) - 3^2}{4^2} = \dfrac{4 - 9}{16} = -\dfrac{5}{16}$

$n = 3 \quad a_3 = \dfrac{2(3) - 3^3}{4^3} = \dfrac{6 - 27}{64} = = -\dfrac{21}{64}$

$n = 4 \quad a_4 = \dfrac{2(4) - 3^4}{4^4} = \dfrac{8 - 81}{254} = -\dfrac{73}{254}$

$n = 5 \quad a_5 = \dfrac{2(5) - 3^5}{4^5} = \dfrac{10 - 243}{1024} = -\dfrac{233}{1024}$

$n = 10 \quad a_{10} = \dfrac{2(10) - 3^{10}}{4^{10}} = \dfrac{20 - 59049}{1048576}$

$= -\dfrac{59029}{1048576}$

$n = 100 \quad a_{100} = \dfrac{2(100) - 3^{100}}{4^{100}}$

B. The Fibonacci sequence is a classic example of a *recursive* sequence. The Fibonacci sequence is given by
$F(n) = F(n - 1) + F(n - 2)$, where $F(1) = F(2) = 1$.
That is, the current term is the sum of the values of the two previous terms. The Fibonacci sequence also has an explicit formula.

2. Find the first twelve terms of the Fibonacci sequence.

$F(1) = 1$
$F(2) = 1$
$F(3) = F(2) + F(1) = 1 + 1 = 2$
$F(4) = F(3) + F(2) = 2 + 1 = 3$
$F(5) = F(4) + F(3) = 3 + 2 = 5$
$F(6) = F(5) + F(4) = 5 + 3 = 8$
$F(7) = F(6) + F(5) = 8 + 5 = 13$
$F(8) = F(7) + F(6) = 13 + 8 = 21$
$F(9) = F(8) + F(7) = 21 + 13 = 34$
$F(10) = F(9) + F(8) = 34 + 21 = 55$
$F(11) = F(10) + F(9) = 55 + 34 = 89$
$F(12) = F(11) + F(10) = 89 + 55 = 144$

3. Find the first ten terms of the recursive sequence defined by
$f(n) = 3 \cdot f(n-1) - f(n-2)$ with $f(1) = 2$ and $f(2) = 3$

$f(1) = 2$
$f(2) = 3$
$f(3) = 3 \cdot f(2) - f(1) = 3 \cdot 3 - 2 = 7$
$f(4) = 3 \cdot f(3) - f(2) = 3 \cdot 7 - 3 = 18$
$f(5) = 3 \cdot f(4) - f(3) = 3 \cdot 18 - 7 = 47$
$f(6) = 3 \cdot f(5) - f(4) = 3 \cdot 47 - 18 = 123$
$f(7) = 3 \cdot f(6) - f(5) = 3 \cdot 123 - 47 = 322$
$f(8) = 3 \cdot f(7) - f(6) = 3 \cdot 322 - 123 = 843$
$f(9) = 3 \cdot f(8) - f(7) = 3 \cdot 843 - 322 = 2207$
$f(10) = 3 \cdot f(9) - f(8) = 3 \cdot 2207 - 843 = 5778$

C. Finding the pattern in a sequence of numbers is a very important concept in mathematics as well as the physical and social sciences. A finite number of terms do not *uniquely* determine a sequence, however in the next problems we are interested in finding an *obvious sequence* whose first few terms agree with the given ones.

4. Find the nth term of the sequence whose first terms are given below.

   a) $\dfrac{2}{3}, \dfrac{8}{3}, \dfrac{32}{3}, \dfrac{128}{3}, \dfrac{512}{3}$

Some facts appear after observing the sequence.
1. The denominator is 3.
2. The numerators are all powers of 2, so we look at the exponent to finish the formula.
$2 = 2^1, \ 8 = 2^3, \ 32 = 2^5, \ 128 = 2^7$
Now find how the exponents 1, 3, 5, 7 are related to the index of the terms 1, 2, 3, 4
So the exponents are $2n - 1$.
Putting it all together we get:
$$a_n = \frac{2^{2n-1}}{3}$$

b) 2, 7, 8, 13, 14, 19

Some facts appear after observing the sequence.
1. The odd number terms are even, the even number terms are odd.
2. The pattern is differ by 5, differ by 1, differ by 5, differ by 1....

The second observation helps a lot, the difference between each consecutive odd terms is 6, the difference between consecutive even terns is 6. This leads to idea that this sequence is $3n + $????
Since 2, 8, and 14 are all $3n - 1$ and 7, 13, and 19 are all $3n + 1$ we just need to get the signs correct. After some trial an error, you should be able to come up with the following formula

$$a_n = 3n + (-1)^n$$

c) $\dfrac{2}{3}, \dfrac{4}{9}, \dfrac{6}{27}, \dfrac{8}{81}$

The numerator and the denominator have different types of sequences going.

First the numerators are all multiples of 2, and after some observation the formula is $2n$. The denominator is 3 to some power and after some trial and error the formula is given by 3^n.

$$a_n = \dfrac{2n}{3^n}$$

Section 9.2 Arithmetic and Geometric Sequences

Key Ideas
A. Arithmetic sequences.
B. Geometric Sequences.

A. An **arithmetic sequence** is a sequence of the form $a, a + d, a + 2d, a + 3d, \ldots$ The number a is the **first term** and the number d is called the **common difference**. Any two consecutive terms of a arithmetic sequence differ by d. The nth term of the arithmetic sequence is given by the formula: $a_n = a + (n - 1)d$.

1. Write the first five terms of a sequence where 9 is the first number and 5 is the common difference.

   $a_n = a + (n - 1)d$
   $a_1 = 9$
   $a_2 = 9 + (2 - 1)5 = 9 + 5 = 14$
   $a_3 = 9 + (3 - 1)5 = 9 + 10 = 19$
   $a_4 = 9 + (4 - 1)5 = 9 + 15 = 24$
   $a_5 = 9 + (5 - 1)5 = 9 + 20 = 29$

2. Given the first 4 terms of a sequence are 6, 13, 20, 27. Find a and d.

   a is the first term, so $a = 6$
   d is the difference between any two terms, so
   $d = 13 - 6 = 7$

   $a_n = 6 + (n - 1)7$

3. The 5th term of an arithmetic sequence is 34 and the 8th term is 43. Find a and d.

   Substitute for n in the formula $a_n = a + (n - 1)d$
   we get:
   $a_5 = 34 = a + (5 - 1)d$
   $a_8 = 43 = a + (8 - 1)d$
   Solve the system
   $34 = a + 4d$
   $43 = a + 7d$ Subtracting
   $9 = 3d$ $d = 3$
   Substitute for d and solve for a.
   So $a = 34 - 4d = 34 - 4(3) = 22$

   $a_n = 22 + (n - 1)3$

229

B. A **geometric sequence** is a sequence of the form $a, ar, ar^2, ar^3, ar^4, \ldots$
The number a is the **first term** and r is the **common ratio**. Any two consecutive terms of a arithmetic sequence differ by d. The nth term of the sequence is given by the formula: $a_n = ar^{n-1}$.

4. Write the first five terms of a sequence where 9 is the first number and 4 is the common ratio.

$a_n = ar^{n-1}$
$a_1 = 9 \cdot 4^{1-1} = 9$
$a_2 = 9 \cdot 4^{2-1} = 9 \cdot 4 = 36$
$a_3 = 9 \cdot 4^{3-1} = 9 \cdot 16 = 144$
$a_4 = 9 \cdot 4^{4-1} = 9 \cdot 64 = 576$
$a_5 = 9 \cdot 4^{5-1} = 9 \cdot 256 = 2304$

5. The 3rd term of an geometric sequence is 6 and the 5th term is $\frac{243}{2}$. Find the first term and the common ratio.

Substitute in the value for a_n and n into $a_n = ar^{n-1}$ and solve.

$a_3 = 6 = ar^{3-1} = ar^2$

$a_5 = \frac{243}{2} = ar^{5-1} = ar^4$

Divide a_3 by a_5 and solve for r

$\dfrac{a_5}{a_3} = \dfrac{ar^4}{ar^2} = r^2 = \dfrac{\frac{243}{2}}{6}$

$r^2 = \dfrac{81}{4}$ so $r = \dfrac{9}{2}$

Substituting for r and solve for a.

$a_3 = 6 = a\left(\dfrac{9}{2}\right)^2$

$6 = \dfrac{81}{4}a$

$a = \dfrac{8}{27}$

Section 9.3 Series

Key Ideas
A. Sigma notation and finite sums.
B. Partial sums and finite series.

A. Sigma notation is a way of expressing the sum of the first n terms of the sequence.
$$\sum_{k=1}^{n} a_k = a_1 + a_2 + a_3 + a_4 + \ldots + a_n$$
This is read as "the sum of a_k from $k = 1$ to $k = n$."

The letter k is called the **index of summation** or the **summation variable** and the idea is to let k take on the values 1, 2, 3, 4, ..., n. Any letter can be used for the index of summation and the index need not start at 1.

Sums have the following properties:
Let $a_1, a_2, a_3, a_4, \ldots$ and $b_1, b_2, b_3, b_4, \ldots$ be sequences. Then for every positive integer n and any constant c,

1. $\sum_{k=1}^{n}(a_k + b_k) = \left(\sum_{k=1}^{n} a_k\right) + \left(\sum_{k=1}^{n} b_k\right)$

2. $\sum_{k=1}^{n}(a_k - b_k) = \left(\sum_{k=1}^{n} a_k\right) - \left(\sum_{k=1}^{n} b_k\right)$

3. $\sum_{k=1}^{n}(c a_k) = c\left(\sum_{k=1}^{n} a_k\right)$

1. Find the following sums.

$$\sum_{k=1}^{4} \frac{1}{k^2}$$

$$\sum_{k=1}^{4} \frac{1}{k^2} = \frac{1}{1^2} + \frac{1}{2^2} + \frac{1}{3^2} + \frac{1}{4^2}$$
$$= 1 + \frac{1}{4} + \frac{1}{9} + \frac{1}{16} = \frac{205}{144}$$

2. Find the following sums.

$$\sum_{j=3}^{7} (-1)^j j^3$$

$$\sum_{j=3}^{7} (-1)^j j^3 = (-1)^3 3^3 + (-1)^4 4^3 + (-1)^5 5^3 +$$
$$(-1)^6 6^3 + (-1)^7 7^3$$
$$= -27 + 64 - 125 + 216 - 343 = -215$$

3. Write the following sum in sigma notation.
$$\frac{2}{3} + \frac{3}{4} + \frac{4}{5} + \frac{5}{6} + \frac{6}{7}.$$

We find the sequence associate to the terms:
$$\frac{2}{3}, \frac{3}{4}, \frac{4}{5}, \frac{5}{6}, \frac{6}{7}$$
Both the numerator and denominator are arithmetic sequences.
The numerator is the sequence $k + 1$ while the denominator is the sequence $k + 2$.

$$\sum_{k=1}^{6} \frac{k+1}{k+2}$$

B. Let $a_1, a_2, a_3, \ldots$ be a sequence. A sum of the form $S = a_1 + a_2 + a_3 + a_4 + \ldots + a_N$ is called a **series**. The number a_1 is called the **first term** of the series, a_2 the **second term**, and so on. The (finite) number S that the series adds to is called the **sum** of the series. For the series $S = a_1 + a_2 + a_3 + a_4 + \ldots + a_N$ the **partial sums** are

$$S_1 = a_1$$
$$S_2 = a_1 + a_2$$
$$S_3 = a_1 + a_2 + a_3$$
$$S_4 = a_1 + a_2 + a_3 + a_4$$
$$\vdots$$
$$S_n = a_1 + a_2 + a_3 + a_4 + \ldots + a_n$$
$$\vdots$$

S_1 is called the **first partial sum**, S_2 the **second partial sum**, and so on. S_n is called the **nth partial sum**. The sequence $S_1, S_2, S_3, S_4, \ldots, S_n, \ldots$ is called the **sequence of partial sums**. The sequence of partial sums is useful in detecting a pattern to find the sum.

4. Find the first six partial sums of the sum
$$\sum_{n=1}^{75} \frac{2n}{n+3}.$$

$$S_1 = \frac{2}{4} = \frac{1}{2}$$
$$S_2 = \frac{2}{4} + \frac{4}{5} = \frac{13}{10}$$
$$S_3 = \frac{2}{4} + \frac{4}{5} + \frac{6}{6} = \frac{23}{10}$$
$$S_4 = \frac{2}{4} + \frac{4}{5} + \frac{6}{6} + \frac{8}{7} = \frac{241}{70}$$
$$S_5 = \frac{2}{4} + \frac{4}{5} + \frac{6}{6} + \frac{8}{7} + \frac{10}{8} = \frac{657}{140}$$
$$S_6 = \frac{2}{4} + \frac{4}{5} + \frac{6}{6} + \frac{8}{7} + \frac{10}{8} + \frac{12}{9} = \frac{2531}{420}$$

5. Find a formula for the nth partial sum of the series and find the sum.
$$\sum_{k=1}^{26} \left(\frac{1}{2}\right)^k$$

Start by finding the first several partial sums until we find a pattern.
$$S_1 = \frac{1}{2} = \frac{1}{2}$$
$$S_2 = \frac{1}{2} + \frac{1}{4} = \frac{3}{4}$$
$$S_3 = \frac{1}{2} + \frac{1}{4} + \frac{1}{8} = \frac{7}{8}$$
$$S_4 = \frac{1}{2} + \frac{1}{4} + \frac{1}{8} + \frac{1}{16} = \frac{15}{16}$$

The pattern is: $S_n = \dfrac{2^n - 1}{2^n}$

$$S_{26} = \frac{2^{26} - 1}{2^{26}} = \frac{67108863}{67108864} \approx 0.999999985$$

6. Find a formula for the nth partial sum of the series and find the sum.
$$\sum_{k=1}^{78} \left(\frac{1}{k+1} - \frac{1}{k+3}\right)$$

Start by finding the first several partial sums until we find a pattern.
$$S_1 = \left(\frac{1}{2} - \frac{1}{4}\right)$$
$$S_2 = \left(\frac{1}{2} - \frac{1}{4}\right) + \left(\frac{1}{3} - \frac{1}{5}\right)$$
$$S_3 = \left(\frac{1}{2} - \frac{1}{4}\right) + \left(\frac{1}{3} - \frac{1}{5}\right) + \left(\frac{1}{4} - \frac{1}{6}\right)$$
$$= \left(\frac{1}{2} + \frac{1}{3}\right) - \left(\frac{1}{5} + \frac{1}{6}\right)$$

At this point a pattern is starting to emerge.
$$S_4 = \left(\frac{1}{2} - \frac{1}{4}\right) + \left(\frac{1}{3} - \frac{1}{5}\right) + \left(\frac{1}{4} - \frac{1}{6}\right) + \left(\frac{1}{5} - \frac{1}{7}\right)$$
$$= \left(\frac{1}{2} + \frac{1}{3}\right) - \left(\frac{1}{6} + \frac{1}{7}\right)$$

This is a telescoping sum.
$$S_n = \left(\frac{1}{2} + \frac{1}{3}\right) - \left(\frac{1}{n+2} + \frac{1}{n+3}\right)$$

So
$$S_{78} = \left(\frac{1}{2} + \frac{1}{3}\right) - \left(\frac{1}{80} + \frac{1}{81}\right) = \frac{5239}{6480}$$

Section 9.4 Arithmetic and Geometric Series

Key Ideas
A. Arithmetic series.
B. Geometric series.

A. An **arithmetic series** is a series whose terms form an arithmetic sequence and are of the form $a_k = a + (k-1)d$. The sum of the first n terms of the series is

$$S_n = \sum_{k=1}^{n} [a + (k-1)d] = a + (a+d) + (a+2d) + \ldots + (a + (n-1)d)$$

is given by $S_n = \dfrac{n}{2}[2a + (n-1)d] = n\left(\dfrac{a + a_n}{2}\right)$

1. Find the sum of the first 40 positive multiples of 3.

   This is the sequence 3, 6, 9, 12, ... where $a = 3$ and $d = 3$.
   So $a_n = 3 + (40-1)3 = 120$

   $$S_{40} = \sum_{k=1}^{40} a_k = 40\left(\dfrac{3 + 120}{2}\right) = 2460$$

2. Find the sum, S_n, of the arithmetic sequence where $a = -2$, $d = 8$, and $n = 5$.

   Using $S_n = \dfrac{n}{2}[2a + (n-1)d]$

   $S_5 = \dfrac{5}{2}[2(-2) + (5-1)8] = \dfrac{5}{2}(-4 + 32)$

   $= \dfrac{5}{2}(28) = 70$

3. Find the sum, S_n, of the arithmetic sequence where $a = 0.31$ and $a_{10} = 8.35$.

   This time use the formula $S_n = n\left(\dfrac{a + a_n}{2}\right)$ with $n = 10$ to find the sum.

   $S_{10} = 10\left(\dfrac{0.31 + 8.35}{2}\right) = 10 \cdot \dfrac{8.66}{2} = 43.3$

B. A **geometric series** is a series whose terms form a geometric sequence whose and are of the form $a_k = ar^k$. The sum of the first n terms of the series is

$$S_n = \sum_{k=0}^{n} ar^k = a + ar + ar^2 + ar^2 + \ldots + ar^{n-1} \quad (r \neq 1)$$

is given by $S_n = a\left(\dfrac{1-r^n}{1-r}\right)$.

4. Find the sum, S_n, of the first 20 terms of the geometric series where $a = 5.6$ and $r = 2$.

Using the formula $S_n = a\left(\dfrac{1-r^n}{1-r}\right)$ and substituting the values $n = 20$, $a = 5.6$ and $r = 2$.

$$S_{20} = (5.6)\left(\dfrac{1-2^{20}}{1-2}\right) = 5.6 \cdot 1048575 = 5872020$$

5. Find the sum of $\displaystyle\sum_{k=6}^{15} (0.3)^k$.

To use the formula we *can* first express the series with the index starting at 0.

$$\sum_{k=6}^{15} (0.3)^k = \sum_{k=0}^{15} (0.3)^k - \sum_{k=0}^{5} (0.3)^k$$

a is found when $k = 0$, so $a = (0.3)^0 = 1$

$$\sum_{k=0}^{15} (0.3)^k = 1\left(\dfrac{1-(0.3)^{15}}{1-(0.3)}\right) = \dfrac{0.999999985}{0.7}$$
$$= 14.2857$$

$$\sum_{k=0}^{5} (0.3)^k = 1\left(\dfrac{1-(0.3)^5}{1-(0.3)}\right) = \dfrac{0.99757}{0.7} = 1.4251$$

$$\sum_{k=6}^{15} (0.3)^k = 14.2857 - 1.4251 = 12.8606$$

Section 9.5 Annuities and Installment Buying

Key Ideas
A. Annuities – future value
B. Present Value of annuity and installment payments.

A. An **annuity** is a sum of money that is paid into an account (or fund) in regular equal intervals. The amount of an annuity, A_f, is the sum of all the individual payments from the beginning until the last payment is made including interest. This is a finite geometric series and results in the formula $A_f = R \dfrac{(1+i)^n - 1}{i}$, where R is the amount of the regular payments, i is the interest per compounding period, and n is the number of payments. Since the value of the money is paid in the future, this is called *future value*.

1. New parents put away $25 a month into a saving account paying 8% compounded monthly. How much will accumulate in this account in 16 years?

   Apply $A_f = R \dfrac{(1+i)^n - 1}{i}$
   with $R = \$25$; $n = 16 \cdot 12 = 192$ payments;
   $i = \dfrac{0.08}{12} = 0.00667$

   $A_f = (25) \dfrac{(1.00667)^{192} - 1}{0.00667} = \$9680.99$

2. A small company decides to save money to purchase a coping machine that cost $9,500. How much do they need to deposit in an account that pays 6% compounded monthly if they need to save this money over a 9 month period?

   Substitute the values $A_f = 9500$, $n = 9$, and $i = \dfrac{0.06}{12} = 0.005$ into $A_f = R \dfrac{(1+i)^n - 1}{i}$ and solve for R.

   $9500 = R \dfrac{(1.005)^9 - 1}{0.005}$

   $9500 = R \dfrac{0.045910579}{0.005} = 9.182115828 R$

   $R = \$1034.62$ rounding up to the next cent.

B. The **present value** is today's value of an amount to be paid in n compounded periods in the future. $PV = A(1+i)^{-n}$, where A is the amount to be paid in the future, i is the interest per compounding period, and n is the number of compounding periods in the future. The present value A_p of an annuity consisting of n regular equal payments of size R with interest rate i per compounding period is given by $A_p = R \dfrac{1-(1+i)^{-n}}{i}$. This formula can be solved for R and the amount of a regular payment on a installment loan is $R = \dfrac{iA_p}{1-(1+i)^{-n}}$. Although it is not possible to algebraically solve the formula for i, it is possible to use a graphing devices to find i.

3. Suppose you plan to purchase a Compact Disc Player from a department store for $300. If you put the $300 on a charge card at 19.8% interest compounded monthly and pay it off in 10 equal payments, how much will each payment be?

   Substitute the values $A_p = 300$, $n = 10$, and
   $$i = \frac{0.198}{12} = 0.0165 \text{ into } R = \frac{iA_p}{1-(1+i)^{-n}} \text{ and}$$
   calculate R.
   $$R = \frac{0.0165 \cdot 300}{1-(1.0165)^{-10}} = \frac{4.95}{0.150963872} = \$32.79 \text{ again}$$
   round up to the next cent.

4. Suppose you want to purchase a car and have $125.00 a month that you could spend on payments for the next 4 years. If the current interest rate on used cars is 11.4%, how much can the car cost?

   Substitute the values $R = 125$, $n = 4 \cdot 12 = 48$, and
   $$i = \frac{0.114}{12} = 0.0095 \text{ into } R = \frac{iA_p}{1-(1+i)^{-n}} \text{ and}$$
   solve for A_p.
   $$125 = \frac{0.0095 A_p}{1-(1.0095)^{-48}} = \frac{0.0095 A_p}{0.364820486} =$$
   $0.026040204 A_p$
   $A_p = \$4800.26$ this time round down to the next cent.

Section 9.6 Infinite Geometric Series

Key Ideas
A. The sum of an infinite geometric series.

A. A series of the form $a + ar + ar^2 + ar^2 + \ldots + ar^{n-1} + \ldots$ is an **infinite series**. From section 9.4, the nth partial sum of such a series is $S_n = a\left(\dfrac{1 - r^n}{1 - r}\right)$ $r \neq 1$. If $|r| < 1$, then r^n gets close to 0 as n gets large, so S_n gets close to $\dfrac{a}{1-r}$. That is the if $|r| < 1$, then the sum of $a + ar + ar^2 + ar^2 + \ldots + ar^{n-1} + \ldots = \dfrac{a}{1-r}$

1. Find the sum of the infinite geometric series
$\dfrac{4}{3} + \dfrac{4}{9} + \dfrac{4}{27} + \dfrac{4}{81} + \ldots$

We need to find a and r and then substitute these values into the formula. a is the first term, so $a = \dfrac{4}{3}$. r is the common ratio and can be found by dividing to consecutive terms. So $r = \dfrac{\frac{4}{9}}{\frac{4}{3}} = \dfrac{1}{3}$

$S = \dfrac{a}{1-r} = \dfrac{\frac{4}{3}}{1 - \frac{1}{3}} = 2$

2. Find the sum of the infinite geometric series
$27 + 18 + 12 + 8 + \ldots$

Find a and r and then substitute these values into the formula. a is the first term, so $a = 27$. r is the common ratio, found by dividing to consecutive terms. So $r = \dfrac{18}{27} = \dfrac{2}{3}$

$S = \dfrac{a}{1-r} = \dfrac{27}{1 - \frac{2}{3}} = 81$

3. Find the sum of the infinite geometric series
$$\frac{3}{\sqrt{5}} - \frac{6}{5} + \frac{12}{5\sqrt{5}} - \frac{24}{25} + \ldots$$

Find a and r and then substitute these values into the formula.

$$a = \frac{3}{\sqrt{5}}$$

$$r = \frac{-\frac{6}{5}}{\frac{3}{\sqrt{5}}} = -\frac{2}{\sqrt{5}}$$

$$S = \frac{a}{1-r} = \frac{\frac{3}{\sqrt{5}}}{1 + \frac{2}{\sqrt{5}}} = \frac{\frac{3}{\sqrt{5}}}{\frac{\sqrt{5}+2}{\sqrt{5}}}$$

$$= \frac{3}{\sqrt{5}+2} = \frac{3}{(\sqrt{5}+2)} \cdot \frac{(\sqrt{5}-2)}{(\sqrt{5}-2)}$$

$$= \frac{3(\sqrt{5}-2)}{5-4} = 3\sqrt{5} - 6$$

4. Express the repeating decimal 0.090909...as a fraction.

$0.09090909\ldots = 0.09 + 0.0009 + 0.000009 + \ldots$

$$= \frac{9}{100} + \frac{9}{100} \cdot \frac{1}{100} + \frac{9}{100} \cdot \frac{1}{100^2} + \ldots$$

So $a = \frac{9}{100}$ and $r = \frac{1}{100}$

$$\frac{a}{1-r} = \frac{\frac{9}{100}}{1 - \frac{1}{100}} = \frac{9}{99} = \frac{1}{11}$$

5. Express the repeating decimal 0.345454545... as a fraction.

0.345454545... = 0.3 + 0.045 + 0.00045 + 0.000045 ...

The geometric series involved is

0.045 + 0.00045 + 0.000045 ...

$$= \frac{45}{1000} + \frac{45}{100000} + \frac{45}{10000000} + \cdots$$

$$= \frac{45}{1000} + \frac{45}{1000} \cdot \frac{1}{100} + \frac{45}{1000} \cdot \frac{1}{100^2} + \cdots$$

So $a = \frac{45}{1000}$ and $r = \frac{1}{100}$

$$\frac{a}{1-r} = \frac{\frac{45}{1000}}{1 - \frac{1}{100}} = \frac{45}{990} = \frac{1}{22}$$

$$0.345454545... = \frac{3}{10} + \frac{1}{22} = \frac{38}{110} = \frac{19}{55}$$

Section 9.7 Mathematical Induction

Key Ideas
A. Principle of mathematical induction.

A. Mathematical induction is a type of proof to show a statement depending on n is true for all natural numbers n. Let $P(n)$ be a statement depending on n. The principle of mathematical induction states that suppose that the following two conditions are true
   I. $P(1)$ is true.
   II. For every natural number k, if $P(k)$ is true, then $P(k + 1)$ is true.
Then $P(n)$ is true for all natural numbers n.
Proof by mathematical induction involves the following steps.
1. Basis of induction: *Prove* that the statement, $P(n)$, is true for some initial value, usually this value is 1.
2. Inductive hypothesis: *Assume* that the statement $P(k)$ is true for some value of k greater than or equal to the initial value.
3. The inductive step: *Prove* that the statement $P(k)$ is true implies that $P(k + 1)$ is true.
4. Conclusion: Since the two conditions have been shown true, by the principle of mathematical induction, the statement is true for all natural numbers.

1. Prove that $n^3 - n + 3$ is divisible by 3 for all natural numbers n.

   Basis of induction : $n = 1$

   $(1)^3 - 1 + 3 = 3$ which is divisible by 3.

   Inductive hypothesis : Assume that the statement is true for some $k \geq 1$, that is, assume that $k^3 - k + 3$ is divisible by 3.

   Inductive step.
   Consider: $(k + 1)^3 - (k + 1) + 3$
   $= (k^3 + 3k^2 + 3k + 1) - k - 1 + 3$
   $= (k^3 - k + 3) + (3k^2 + 3k + 3)$
   $= (k^3 - k + 3) + 3(k^2 + k + 1)$

   Since $k^3 - k + 3$ is divisible by 3 (by the inductive hypotheses).
   And since $3(k^2 + k + 1)$ is clearly divisible by 3, the sum.
   $(k^3 - k + 3) + 3(k^2 + k + 1) = (k + 1)^3 - (k + 1) + 3$ is divisible by 3.

   Conclusion. Since the formula is true for $n = 1$, and $k^3 - k + 3$ is divisible by 3 implies $(k + 1)^3 - (k + 1) + 3$ is divisible by 3, by the principle of mathematical induction, the statement is true for all natural numbers n.

2. Let $a_n = a_{n-1} + 2a_{n-2} - 6$ with $a_1 = 4$ and $a_2 = 5$. Find the first 5 terms of the sequence and prove that $a_n = 2^{n-1} + 3$.

$a_1 = 4; a_2 = 5;$
$a_3 = a_2 + 2a_1 - 6 = 5 + 2(4) - 6 = 7$
$a_4 = a_3 + 2a_2 - 6 = 7 + 2(5) - 6 = 11$
$a_5 = a_4 + 2a_3 - 6 = 11 + 2(7) - 6 = 19$

Basis of induction :

Show that the statement is true for $n = 1$ and $n = 2$. (Note both these cases are need because a_n is based on the terms a_{n-1} and a_{n-2}.)

$a_1 = 2^{1-1} + 3 = 2^0 + 3 = 1 + 3 = 4$
$a_2 = 2^{2-1} + 3 = 2^1 + 3 = 2 + 3 = 5$

Inductive hypotheses :

Assume that the statement is true for $k \geq 2$ and for all values $m \leq k$.
Note we had to start at $k \geq 2$ because a_2 is not given by the formula $a_n = a_{n-1} + 2a_{n-2} - 6$.
Also note since both the terms a_{n-1} and a_{n-2} are used, we need an inductive hypothese both terms. Here we assume that :

$a_k = a_{k-1} + 2a_{k-2} - 6 = 2^{k-1} + 3$

is true for some k and $k - 1$.

Inductive step.

We <u>need</u> to show that

$a_{k+1} = a_{k+1-1} + 2a_{k+1-2} - 6 = 2^{k+1-1} + 3$

or: $a_{k+1} = a_k + 2a_{k-1} - 6 = 2^k + 3$

By definition,

$a_{k+1} = a_k + 2a_{k-1} - 6$

Substituting for a_k and a_{k-1} we get

$a_{k+1} = (2^{k-1} + 3) + 2(2^{k-2} + 3) - 6$
$= 2^{k-1} + 3 + 2^{k-1} + 6 - 6$
$= 2^{k-1} + 2^{k-1} + 3$
$= 2^k + 3$

So the statement is for $k + 1$.

Conclusion.

By the principles of mathematical induction,
$a_{n-1} + 2a_{n-2} - 6 = a_n = 2^{n-1} + 3$ (with $a_1 = 4$ and $a_2 = 5$), is true for all natural numbers n.

Section 9.8 The Binomial Theorem

Key Ideas
A. Pascal's triangle.
B. Binomial coefficients of Pascal's triangle.

A. An expression of the form $a + b$ is called a **binomial**. When $(a + b)^n$ is expanded, the following patterns emerge.
  1. There are $n + 1$ terms, The first being $a^n b^0$ and the last $a^0 b^n$.
  2. The **exponents of a decrease by 1** from term to term while the **exponents of b increase by 1**.
  3. The sum of the exponents of a and b in each term is n.

Pascal's triangle is the following triangular array.

$(a + b)^0$ 1
$(a + b)^1$ 1   1
$(a + b)^2$ 1   2   1
$(a + b)^3$ 1   3   3   1
$(a + b)^4$ 1   4   6   4   1
$(a + b)^5$ 1   5   10   10   5   1
$(a + b)^6$ 1   6   15   20   15   6   1

Every entry (other than a 1) is the sum of the two entries diagonally above it.

1. Use Pascal's triangle above to expand $(3 + 2x)^5$.

$(a + b)^5 = a^5 + 5a^4 b^1 + 10a^3 b^2 + 10a^2 b^3 + 5a^1 b^4 + b^5$

$(3 + 2x)^5 = (3)^5 + 5(3)^4 (2x)^1 + 10(3)^3 (2x)^2 +$
$10(3)^2 (2x)^3 + 5(3)^1 (2x)^4 + (2x)^5$

$= 243 + 90x + 1080x^2 + 720x^3 + 80x^4 + 32x^5$

2. Use Pascal's triangle above to expand $(3x - 5y)^3$.

$(a + b)^3 = a^3 + 3a^2 b^1 + 3a^1 b^2 + b^3$

$(3x - 5y)^3 = (3x)^3 + 3(3x)^2 (-5y)^1 + 3(3x)^1 (-5y)^2 +$
$(-5y)^3$

$= 27x^3 - 135x^2 y + 225xy^2 - 125y^3$

B. Pascal's triangle can also be written as

$$\binom{0}{0}$$
$$\binom{1}{0} \quad \binom{1}{1}$$
$$\binom{2}{0} \quad \binom{2}{1} \quad \binom{2}{2}$$
$$\binom{3}{0} \quad \binom{3}{1} \quad \binom{3}{2} \quad \binom{3}{3}$$
$$\binom{4}{0} \quad \binom{4}{1} \quad \binom{4}{2} \quad \binom{4}{3} \quad \binom{4}{4}$$
$$\binom{5}{0} \quad \binom{5}{1} \quad \binom{5}{2} \quad \binom{5}{3} \quad \binom{5}{4} \quad \binom{5}{5}$$
$$\binom{6}{0} \quad \binom{6}{1} \quad \binom{6}{2} \quad \binom{6}{3} \quad \binom{6}{4} \quad \binom{6}{5} \quad \binom{6}{6}$$

For any nonnegative integers r and k with $r \leq k$,

$$\binom{k}{r-1} + \binom{k}{r} = \binom{k+1}{r}$$

The **Binomial theorem** states that $(a+b)^n =$

$$\binom{n}{0}a^n + \binom{n}{1}a^{n-1}b^1 + \binom{n}{2}a^{n-2}b^2 + \ldots + \binom{n}{n-1}a^1 b^{n-1} + \binom{n}{n}b^n$$

The term that contains a^r in the expansion of $(a+b)^n$ is $\binom{n}{r}a^r b^{n-r}$.

3. Evaluate $\binom{4}{0} + \binom{4}{1} + \binom{4}{2} + \binom{4}{3} + \binom{4}{4}$ $\Bigg| \binom{4}{0} + \binom{4}{1} + \binom{4}{2} + \binom{4}{3} + \binom{4}{4}$

   is the expansion of $(1+1)^4$

   $\binom{4}{0} + \binom{4}{1} + \binom{4}{2} + \binom{4}{3} + \binom{4}{4} = (1+1)^4 = 2^4$

   $= 16$

4. Evaluate $\binom{4}{0} - \binom{4}{1} + \binom{4}{2} - \binom{4}{3} + \binom{4}{4}$ $\Bigg| \binom{4}{0} - \binom{4}{1} + \binom{4}{2} - \binom{4}{3} + \binom{4}{4}$

   is the expansion of $(1-1)^4$, so

   $\binom{4}{0} - \binom{4}{1} + \binom{4}{2} - \binom{4}{3} + \binom{4}{4} = (1-1)^4 = 0$

5. Find the first three terms in the expansion of $(4-x)^9$.

$\binom{9}{0}4^9 x^0 - \binom{9}{1}4^8 x^1 + \binom{9}{2}4^7 x^2$

$= 262144 - 589824x + 589824x^2$

6. Find the last three terms in the expansion of $(7+y)^{11}$.

$\binom{11}{9}7^2 y^9 + \binom{11}{10}7^1 y^{10} + \binom{11}{11}7^0 y^{11}$

$= 2695y^9 + 77y^{10} + y^{11}$

7. Find the term containing x^8 in the expansion of $(3-2x)^{15}$.

The term containing a^r is $\binom{n}{r}a^r b^{n-r}$

So the term containing x^8 is $\binom{15}{8}(2x)^8 \, 3^{15-8}$

$= 6435 \cdot 2^8 \cdot x^8 \cdot 3^7 = 3602776320 x^8$